Anthropometry

Today, human factors and ergonomics professionals worldwide contribute to the design and evaluation of tasks, jobs, products, environments, and systems in order to make them compatible with the needs, abilities, and limitations of people. By understanding anthropometry, professionals can ensure that our home and working environments are comfortable and designed with the human in mind. This book aims to show how an understanding of anthropometrics can influence workspace design, ergonomics in the office, ergonomics in the home, and health and safety at work.

This book discusses the measurement of the human body and human variability. Anthropometry may seem to be relatively simple but the reality is that it focuses on very sophisticated aspects of how to make the products tailor-made to suit specific requirements. As a study, it is useful for a variety of purposes such as workspace design, ergonomics in the office, ergonomics in the home, and health and safety at work. These eleven chapters investigate anthropometrics and bridge the gap between theory and practice. Each chapter is supported by tables, charts, and illustrations, and a wide list of bibliographic references. The reader will develop new insights into the principles and practice of anthropometrics with this book bringing the topic right up to date.

Anthropometry: Human Body Measurements and How to Use Them will be of interest to students, graduates, teachers, researchers, and general workers in industrial design, ergonomics, rehabilitation, safety, and health.

Beata Mrugalska is an Associate Professor and Head of the Division of Applied Ergonomics, Institute of Safety and Quality Engineering, Faculty of Management Engineering, Poznan University of Technology in Poland. She holds an MSc (2001) in Management and Marketing from the Faculty of Mechanical Engineering and Management at the Poznan University of Technology, and a PhD (2009) in Machine Construction and Operation from the Faculty of Computer Science and Management at the Poznań University of Technology. She was awarded DSc degree in Mechanical Engineering by the Faculty of Mechanical Engineering of the Poznan University of Technology, Poland (2019). Since 2018, she has been a board member of the Center for Registration of European Ergonomists (CREE).

Waldemar Karwowski is Pegasus Professor and Chairman, Department of Industrial Engineering and Management Systems and Executive Director, Institute for Advanced Systems Engineering, University of Central Florida, Orlando, USA. He holds an MS (1978) in Production Engineering and Management from the Technical University of Wroclaw, Poland, and a PhD (1982) in Industrial Engineering from Texas Tech University. He was awarded DSc degree in Management Science by the State Institute for Organization and Management in Industry, Poland (2004). He is the Past President of the International Ergonomics Association (2000–2003), and Past President of the Human Factors and Ergonomics Society (2007). Dr. Karwowski served on the Committee on Human Factors/Human Systems Integration, National Research Council, the National Academies, USA (2007–2011).

Body of Knowledge in Human Factors and Ergonomics

Series Editors:
Waldemar Karwowski, University of Central Florida,
and Beata Mrugalska, Poznan University of Technology

This Focus book series aims to provide the perspective readers-at-large with basic knowledge of human factors and ergonomics that can be useful to students, teachers, academicians, researchers, and practitioners in domains of human activity, across all sectors of industry, business, education, home, leisure, entertainment, and sports. This new Focus series will cover the whole spectrum of topics related to a broadly defined HF/E field. All potential chapters will be selected from the content of the HF/E encyclopedia, built out and arranged into thematic short form Focus titles on the specific aspects of research, practice, and education within this field.

Should you be interested in including your own book in this series, please contact James Hobbs the Editor for Ergonomics/Human Factors and Occupational Health and Safety on james.hobbs@tandf.co.uk.

Anthropometry: Human Body Measurements and How to Use Them
Edited by Beata Mrugalska and Waldemar Karwowski

For more information about this series, please visit: https://www.routledge.com/Body-of-Knowledge-in-Human-Factors-and-Ergonomics/book-series/CRCFBK

Anthropometry
Human Body Measurements and How to Use Them

Edited by
Beata Mrugalska and
Waldemar Karwowski

Routledge
Taylor & Francis Group

NEW YORK AND LONDON

First published 2024
by Routledge
605 Third Avenue, New York, NY 10158

and by Routledge
4 Park Square, Milton Park, Abingdon, Oxon OX14 4RN

Routledge is an imprint of the Taylor & Francis Group, an informa business

ISBN: 978-1-032-58779-0 (hbk)
ISBN: 978-1-032-60585-2 (pbk)
ISBN: 978-1-003-45976-7 (ebk)

DOI: 10.1201/9781003459767

Typeset in Times New Roman
by codeMantra

Contents

List of Contributors   vii
Preface   ix
Glossary of Anthropometric Terms, K. H. E. Kroemer   xi

1 **Anthropometry: The Past, the Present, and the Future**   1
   B. Mrugalska and W. Karwowski

2 **Anthropometry: Definition, Uses, and Methods of Measurements**   14
   R. E. Herron

3 **Keyword: Body Sizes of Americans**   23
   K. H. E. Kroemer

4 **Anthropometric Databases**   33
   R. E. Herron

5 **Engineering Anthropometry**   38
   K. H. E. Kroemer

6 **Anthropometry for Design**   42
   E. Nowak

7 **Ergonomic Workstation Design**   57
   B. Das

8 **Anthropometric Topography**   80
   Z. Li

9 **Anthropometry of Children**   91
   E. Nowak

10   Anthropometry for the Needs of the Elderly   **110**
E. Nowak

11   Anthropometry for the Needs of Rehabilitation   **128**
E. Nowak

Index   147

Contributors

B. Das,
Department of Industrial
 Engineering, Dalhousie
 University,
Halifax, Nova Scotia, Canada

R. E. Herron,
Department of Ergonomics,
 Colorado State University,
Fort Collins, CO

W. Karwowski,
University of Central Florida,
 Orlando, FL

K. H. E. Kroemer,
ISE Department, Virginia Tech,
 Blacksburg, VA

Z. Li,
Department of Industrial
 Engineering, Tsinghua University,
Beijing, China

B. Mrugalska,
Poznan University of Technology,
Poland

E. Nowak,
Institute of Industrial Design,
 Świętojerska,
Warsaw, Poland

Preface

Today, human factors and ergonomics professionals worldwide contribute to the design and evaluation of tasks, jobs, products, environments, and systems in order to make them compatible with the needs, abilities, and limitations of people (IEA 2000). These professionals, who are often certified by the professional certification bodies (e.g. Board of Certification in Professional Ergonomics in the United Sates, or the Centre for Registration of European Ergonomists—CREE in the European Union), promote the human-centered approach to work systems design, testing, and evaluation, which considers the broad set of physical, cognitive, social, organizational, environmental, and other relevant human factors. Such actions should help in the socio-economic development of the world society at large. As noted by the National Academy of Engineering in the United States, in the near future "engineering will expand toward tighter connections between technology and the human experience ..., and ergonomic design of engineered products" (NAE 2004). For example, the International Ergonomics Association works to enhance the public understanding of the meaning of ergonomics, and facilitates making informed decisions about the promotional claims of "ergonomically designed" systems. In order to be able to achieve it, it is required to refer to the measurements of the human body and its parts to be able to fit it to its users.

This book brings new insights into the principles and practice of anthropometrics, which can be useful for a variety of purposes such as workspace design, ergonomics in the office, ergonomics in the home, and health and safety at work. It focuses not only on anthropometrics but also the science behind anthropometrics and joins two sides of a shore: theory and practice.

REFERENCES

IEA, 2000, International Ergonomics Association: www.iea.cc.
KARWOWSKI, W., 2005, Ergonomics and human factors: the paradigms for science, engineering, design, technology, and management of human-compatible systems. *Ergonomics*, 48(5), 436–463.
NAE, 2004, *The Engineer of 2020: Visions of Engineering in the New Century* (Washington, DC: National Academy of Engineering, National Academies Press).

Glossary of Anthropometric Terms, K. H. E. Kroemer

Anterior	In front, toward the front of the body
Anthropometry	Measure of the human body. The term is derived from the Greek words "anthropos," human and "metron," measure
Breadth	Straight-line, point-to-point horizontal measurement running across the body or a body segment
Circumference	Closed measurement following a body contour; hence this measurement is usually not circular
Coronal	In a plane that cuts the boy into fore–aft (anterior–posterior) sections; same as frontal (see Figure 0.1)
Curvature	Point-to-point measurement following a body contour; this measurement is neither closed nor usually circular
Depth	Straight-line, point-to-point horizontal measurement running fore–aft the body
Distal	Away from the center of the body; opposite of proximal
Distance	Straight-line, point-to-point measurement, usually between landmarks of the body
Dorsal	Toward the back or spine; opposite of ventral
Frontal	In a plane that cuts the boy into fore–aft (anterior–posterior) sections; same as coronal
Height	Straight-line, point-to-point vertical measurement
Inferior	Below, toward the bottom; opposite of superior
Lateral	To the side, away from the middle
Medial	In a plane that cuts the boy into left and right halves; same as mid-sagittal
Mid-sagittal	In a plane that cuts the boy into left and right halves; same as medial
Posterior	Behind, toward the back of the body; opposite of anterior
Proximal	Toward or near the center of the body; opposite of distal

Sagittal	In a plane parallel to the medial plane (occasionally used as medial)
Superior	Above, toward the top; opposite of inferior
Reach	Point-to-point measurement following the long axis of an arm or leg
Transverse	In a plane that cuts the boy into upper and lower (superior and inferior) sections
Ventral	Toward the abdomen (occasionally used like anterior)

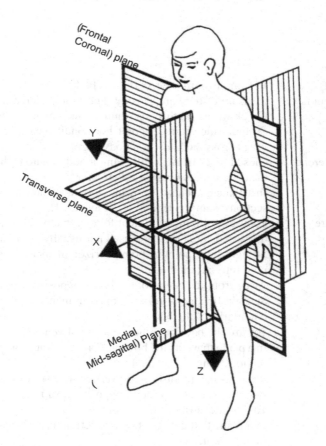

FIGURE 0.1   Anthropometric reference planes.

FURTHER READING

1. ISO 7250-1, Basic human body measurements for technological design—Part 1: Body measurement definitions and landmarks.
2. ISO 15535, General requirements for establishing anthropometric databases.
3. ISO 8559-1, Size designation of clothes—Part 1: Anthropometric definitions for body measurement.
4. ISO19408, Footwear—Sizing—Vocabulary and terminology.
5. ISO 15536-1, Ergonomics—Computer manikins and body templates—Part 1: General requirements.

Anthropometry
The Past, the Present, and the Future

1

B. Mrugalska and W. Karwowski

1.1 HISTORY OF ANTHROPOMETRY

The term "Anthropometry" derives from two Greek words *anthrop* (human) and *metricos* (measurement). Their meaning shows directly that it refers to measurement of people. In practice it focuses on both the process of finding and collecting individual physical characteristics such as body dimensions, body volumes, masses of body segments, center of gravity and inertial properties (Marshall & Summerskill, 2019). Although the human measurements originate in ancient civilizations, its genesis goes back to the 19th century. Then it was used to study of human variation and evolution regarding living and extinct populations by physical anthropologists. The aim of these first studies was to match racial, cultural and psychological attributes with physical properties (Titoria & Sharma, 2022).

The father of anthropometry, also known as a father of criminal identification, is Alphonse Bertillon. He developed a new classification system called

DOI: 10.1201/9781003459767-1

the "anthropometric system" or "judicial anthropometry" which reflected to anthropomorphic measurements and two assumptions:

- bone density is fixed past the age of 20 years,
- human dimensions are highly variable.

In order to elaborate this system he took measure of height, breadth, foot size, length and width of the head, length of the middle finger and the length of the left forearm of criminals. He increased these data by morphological and individual features. Furthermore, he classified them into small, medium, or large, and added frontal and profile photography to each person's file. His approach was implemented in the Paris criminology department to identify unknown personalities and repeat lawbreakers. His anthropometric system called "Bertillonage" started to be used worldwide at the turn of the 19th and 20th centuries and nowadays it is known as a "mug shot" (Helfand, 2019; Laws, 2020).

The earliest anthropologists were only concerned with body measurements in the field of physical anthropology. They emphasized the benefits of such data for industrial design, art, military purposes, medical research and detection of body defects and their correction and forensic identification (Hrdlica, 1920). In the 1930s–1940s an emphasis was put on military anthropological research in both US and Europe. Joining the knowledge of anthropology, psychology, physiology and medicine helped to create a new domain of engineering known as ergonomics or human factors. Since World War II, it was noticed that the well-designed workplaces and equipment increase work performance the efficiency and lower fatigue of the workers. It changed the focus of the anthropometry to be more oriented to deliver a good fit product to a daily user (Zakaria & Gupta, 2019). Nowadays, engineering anthropometry supports mainly industrial design, clothing design, medicine and architecture. These data support design and evaluate – customer, system, equipment, tools, products, work spaces and facilities (Dianat et al., 2018).

1.2 TRADITIONAL ANTHROPOMETRY

Traditional anthropometry refers to human anatomy where collecting measurements of the physical properties (heights, depths and widths) requires to follow the standardized postures where the distance between body landmarks is precisely described (Ma & Ni, 2021). The landmarks are defined as a bony prominence or other physically definable points, which can be located by identifying the bones beneath the skin. They are marked on the body as they are assigned to be endpoints for measurements (Gupta, 2014). All these measurements are noninvasive in spite of the fact that they require direct contact (Ma & Ni, 2021).

Traditional anthropometric measurement methods can be divided into one-dimensional (1D) direct manual measurements and two-dimensional (2D) photogrammetric methods (Dianat et al., 2018). The direct measurements require using cheap and easy tools such as flexible, non- stretchable measuring tapes, skinfold calipers, knee calipers, stadiometers, anthropometers, sitting-height tables, softer ruler and infantometer to measure the recumbent length. Some of these traditional anthropometric tools need calibration to a prescribed schedule and method to ensure the accuracy of the measurement (Bragança et al., 2016; Ma & Ni, 2021). However, the accuracy of the measurement can be sometimes affected by the mood and state of the measurer and the investigated subject (Ma & Ni, 2021). In order to do it correctly, they should be performed by professional examiners. Moreover, multiple measurements should be taken to calculate the means and the means should be used for further investigations (Bragança et al., 2016; Ma & Ni, 2021). However, there might be circumstances in which the measurements might give inaccurate data in spite of fulfilling all these requirements. It can result from serious illness or impossibility to measure deformity or casting of some parts of the body. The problems may also appear when it can be hard to get access to the required reference points for a person in wheelchairs (Sims et al., 2012). Using anthropometric measurement in such situations can give falsely supporting or alarming data (Casadei & Kiel, 2022).

Two-dimensional (2D) photogrammetric methods require multi-camera systems to register relatively simultaneous 2D images of human body from different viewing angles (Yu et al., 2013). However, the acquired images depend on the number of registered images, viewing angle, distortion of a camera lens when taking the images and lighting conditions (Lau & Armstrong 2011; Yu et al., 2013). For example, in the case of certain facial measurements, it is suggested that 2D photogrammetry can support direct measurement. However, it is underlined to carry out a reliability and validation of any measurement methods in anthropometric studies (Lim et al., 2022).

1.3 DIGITAL ANTHROPOMETRY

The investigation of human body using digital capture tools dates back to 1973 when a Lovesey's light sectioning (1966) was used. It is a monophotogrammetric method where the subjects are measured in three dimensions. It consists of contouring the object optically by a series of slits of light of equal size and frequency. In order to perform it, the contour plot is recorded photographically at 90° to the projection axis (Williams, 1977). In the beginning the analysis of data was tremendously time consuming. However, this technology developed rapidly and nowadays it is known as a three-dimensional human body scanner.

A 3D human body scanner, also known as 3D body scanner, operates using diverse technologies such as: optical measurement, computer, image processing and digital signal processing which do not involve physical contact. However, the scanned person frequently wears a closely fit outfit. Therefore, in spite of the fact that the body is not touched, a problem may arise as this method refers to the recognizable image-capture of the semi-nude body. The sensitive personal images and data might be produced, and if not stored safety, can potentially be available across the Internet (Bindahman et al., 2012). Furthermore, such data collection may cause the need of discussing the ethical, social and cultural issues as it involves human body rights (Saaludin et al., 2022).

One of the primary 3D body scanning systems was the Loughborough Anthropometric Shadow Scanner (LASS) which was patented in 1989. In spite of the fact that it was an automated and computerized 3D measurement system, it was based on triangulation. It means that the participant had to stay on a rotating platform, which was turning 360° in the measurement (Jones et al., 1989). Brooke-Wavell et al. (1994) compared it to the traditional measurement methods of anthropometry and they found out that the measurements collected using these two methods were comparable. The statistical differences ($p < 0.05$) were only noticed for women's measurements of neck and chest circumferences, waist width, depth and height, and for men between the measurements of waist depth. It resulted from the positioning of site markers and problems in performing horizontal measurements with the tape measure. The issues of accuracy of landmarks (Kouchi & Mochimaru, 2010), constant body ratio benchmarks (Wang & Chao, 2010) and comparison of the data collected in both methods (Feathers et al., 2004; Han et al., 2010) have been widely discussed. Kouchi and Mochimaru (2010) emphasized that it was related to the error in identifying landmarks. Therefore, they suggested introducing a more clear definition of landmarks in order to decrease these errors and a widely accepted protocol for 3D anthropometry.

Body scanner technologies have developed significantly in the last three decades. Today, it is also possible to find several alternative body scanning systems which use a variety of technologies. There are used diverse application of computer simulations, automated measurement technology and image processing techniques. However, it has to be underlined that the whole (or half) body scanning system takes the advantage of three-dimensional optical scanning from multiple angles and directions within 3–5 seconds. It uses harmless white light to the human body to perform its operations. The accuracy of the collected human body point cloud data is about 0.5 mm what allows us to generate a patch model of precise body parameters. The characteristics of such 3D scanners can be defined as: self-positioning, user-friendly, high precision, high speed, low price, flexibility, multi-purpose, hand-held device, humanization and scalability (https://visbody.com/blog/what-are-the-characteristics-of-3d-body-scanners/).

The rapid development of artificial intelligence and robotics has developed the real-time three-dimensional shape measurement techniques into three categories: structured light, stereo vision and time of flight (Wang, 2020). In the first technique projects a structured light pattern is projected onto the surface of the body from the front and back. It allows calculating a full 3D image using the deformed pattern (Daanen & Haar, 2013; Bragança et al., 2016). In a stereo vision technique the three-dimensional shape is obtained on the basis of the corresponding points between two cameras. Thanks to the knowledge about the fixed relative positioning of these cameras, software matches corresponding points in two flat images, recognizes differences and produces a full 3D point cloud through triangulation (Wang, 2020). In a time-of-flight technique, due to the time delay between the emitted and reflected laser light from the object's surface, we calculate an exact distance (https://www.zivid.com/3d-vision-technology-principles). Contemporary, all these techniques have been widely regarded as breakthrough technological development in both academic and business communities.

Nowadays 3D body scanning systems are commonly used in clothing, animation, ergonomics, medicine and other fields. They enable developing human (face) pattern recognition, fashion designs, protective clothing (for astronauts or divers), special equipment (prosthetics, personalized products) and perform ergonomics research (Raji et al., 2021; Stark et al., 2022). In spite of the fact that the classical apple and pear body shape concepts of man and woman remain useful, the novel 3-D optical measurements will have a potential to change the future in anthropometrics (Minotti et al., 2022).

1.4 ANTHROPOMETRY MANUALS, GUIDELINES AND STANDARDS

Calculating accurate anthropometric data and assuring their appropriate recording is an overall responsibility of a measurer. This person is also in charge for the measurement equipment, suppliers and reporting their malfunctioning to the supervisor. However, in order to avoid unnecessary confusions how to organize and perform such high-quality data collection, it is advisable to follow manuals, guidelines and/or standards. For this aim, UNICEF developed "Manual for anthropometry" (2014) for Multiple Indicator Cluster Surveys (MICS). This manual refers to the resources developed by the World Health Organization (WHO), Action Contre la Faim Canada, and the Food and Nutrition Technical Assistance Project (FANTA). It shows how to accurately measure and weigh children, and provides also general precautions for the measurer and assistant. It is also possible to refer to "Anthropometry Procedure Manual" which is based on National Health and Nutrition Examination Survey (NHANES) (CDC, 2016). NHANES surveys focus on differences in body measurement values instead of technologist and protocol variability and/or measurement error. Therefore, they show the need of training staff to measure and record the survey data with high precision using procedures and calibrated equipment. The NHANES anthropometry data have been used to analyze growth and weight trends in the US population for more than 30 years. The departments of health care services in particular states, also provide health assessment manuals or guidelines in which they focus on anthropometry measurements, i.e. California Department of Health Care Services (2016), Minnesota Department of Health (2022). The accurate anthropometric measurements are paramount to them as the adequate assessment of the health of children and adolescents allows us to choose appropriate treatments and interventions necessary to maintain or improve health. Such documents are used to assess a health during a routine well-patient visit. Looking for a guide to anthropometry, we can also refer to a project Fanta and its results called "The FANTA Guide to Anthropometry" (Cashin & Ott, 2018). Its aim is to perform anthropometric measurement to evaluate and comprehend the nutritional status of individuals and populations. In order to achieve it, it provides data interpretation, measurement protocols and information on equipment selection. It replaces the Anthropometric Indicators Measurement Guide from 2003. It includes the recommendations included in the World Health Organization (WHO) Child Growth Standards and the WHO Growth Reference for children 5–19 years. Furthermore, it discusses a new key malnutrition indicator which is mid-upper arm circumference, and the importance of the nutrition of adolescents and adults in the developing countries. It is worth to underline that this guide concentrates on the anthropometric measurements usually used in low-resource settings. It supports how gather, comprehend, and/or use such data as a part of service provision, surveillance, surveys, monitoring and evaluation program. Another project, proving guidelines to anthropometry measurements, is the INTERGROWTH-21st study which involves health institutions from geographically diverse countries. It does not only show data collection procedures but its scope is extended to the procedures for selection of personnel, measurement equipment and technique, and quality control procedures (2012).

In order to have a comparable data across measures and studies, all anthropometry procedures should be standardized. They need to be internally and externally reliable, within and between the studies. The measurement protocol should be reproducible, comprehensible and acceptable to present and future users. Therefore, in such studies it is worth to rely on international standards. There are several ASTM, EN, IEEE and ISO publications related to anthropometry as depicted in Table 1.1.

TABLE 1.1 Anthropometry in Exemplary ASTM, EN, IEEE and ISO Publications (3rd March 2023)

NUMBER	TITLE OF PUBLICATION
ASTM E3149-18	Standard guide for facial image comparison feature list for morphological analysis
ASTM F1166-21	Standard practice for human engineering design for marine systems, equipment, and facilities
ASTM F3474-20	Standard practice for establishing exoskeleton functional ergonomic parameters and test metrics
ASTM F3518–21	Standard guide for quantitative measures for establishing exoskeleton functional ergonomic parameters and test metrics
EN 547-3:1996+A1:2008	Safety of machinery. Human body measurements – Part 3: Anthropometric data
EN 4730:2018	Aerospace series – Anthropometric dimensioning of aircraft seats
EN 16186-1:2014+A1:2018	Railway applications. Driver's cab Anthropometric data and visibility
IEEE White Paper	IEEE SA 3D Body processing industry connections – comparative analysis of anthropometric methods: Past, present, and future
ISO/TR 7250–2:2010/Amd 1:2013	Basic human body measurements for technological design – Part 2: Statistical summaries of body measurements from national populations – Amendment 1
ISO 7250–3:2015	Basic human body measurements for technological design – Part 3: Worldwide and regional design ranges for use in product standards
ISO/TR 9241–514:2020	Ergonomics of human-system interaction – Part 514: Guidance for the application of anthropometric data in the ISO 9241–500 series
ISO 11226:2000	Ergonomics – Evaluation of static working postures
ISO 11226:2000/Cor 1:2006	Ergonomics – Evaluation of static working postures – Technical Corrigendum 1

(Continued)

TABLE 1.1 (Continued)

NUMBER	TITLE OF PUBLICATION
ISO 11228-1:2021	Ergonomics – Manual handling – Part 1: Lifting, lowering and carrying
ISO 11228-2:2007	Ergonomics – Manual handling – Part 2: Pushing and pulling
ISO 11228-2:2007/Amd 1:2022	Ergonomics – Manual handling – Part 2: Pushing and pulling – Amendment 1
ISO 11228-3:2007	Ergonomics – Manual handling – Part 3: Handling of low loads at high frequency
ISO/TR 12295:2014	Ergonomics – Application document for International Standards on manual handling (ISO 11228-1, ISO 11228-2 and ISO 11228-3) and evaluation of static working postures (ISO 11226)
ISO/TR 12296:2012	Ergonomics – Manual handling of people in the healthcare sector
ISO 13232-3	Motorcycles – Test and analysis procedures for research evaluation of rider crash protective devices fitted to motorcycles – Part 3: Motorcyclist anthropometric impact dummy
ISO 14738:2002	Safety of machinery – Anthropometric requirements for the design of workstations at machinery
ISO 14738:2002/Cor 1:2003	Safety of machinery – Anthropometric requirements for the design of workstations at machinery – Technical Corrigendum 1
ISO 14738:2002/Cor 2:2005	Safety of machinery – Anthropometric requirements for the design of workstations at machinery – Technical Corrigendum 2
ISO 15534-1:2000	Ergonomic design for the safety of machinery – Part 1: Principles for determining the dimensions required for openings for whole-body access into machinery
ISO 15534-2:2000	Ergonomic design for the safety of machinery – Part 2: Principles for determining the dimensions required for access openings
ISO 15534-3:2000	Ergonomic design for the safety of machinery – Part 3: Anthropometric data
ISO 15535:2012	General requirements for establishing anthropometric databases
ISO 15536-1:2005	Ergonomics – Computer manikins and body templates – Part 1: General requirements

(Continued)

TABLE 1.1   (Continued)

NUMBER	TITLE OF PUBLICATION
ISO 15536-2:2007	Ergonomics – Computer manikins and body templates – Part 2: Verification of functions and validation of dimensions for computer manikin systems
ISO 15537:2022	Principles for selecting and using test persons for testing anthropometric aspects of industrial products and designs
ISO/TS 20646:2014	Ergonomics guidelines for the optimization of musculoskeletal workload
ISO 20685-1:2018	3-D scanning methodologies for internationally compatible anthropometric databases – Part 1: Evaluation protocol for body dimensions extracted from 3-D body scans
ISO 20685-2:2015	Ergonomics -3-D scanning methodologies for internationally compatible anthropometric databases – Part 2: Evaluation protocol of surface shape and repeatability of relative landmark positions
ISO/TR 23076:2021	Ergonomics – Recovery model for cyclical industrial work
ISO/TR 23476:2021	Ergonomics – Application of ISO 11226, the ISO 11228 series and ISO/TR 12295 in the agricultural sector
ISO 24553:2023	Ergonomics – Accessible design – Ease of operation
PD CEN/TR 16823:2015	Railway applications. Driver's cab. Background information on anthropometric data
PD CEN/TR 17698:2021	Ergonomics. Demands and availability of anthropometric and strength data of children in Europe

Source: Full text of the above standards can be found on the following websites: https://www.iso.org, http://www.en-standard.eu, https://www.astm.org

As it can be noticed the ASTM standards, formerly standards of American Society for Testing and Materials, European Standards (EN) and ISO publications encompass activities in relation to a wide range of fields and sectors. ISO publications include not only ISO Standards, but also ISO/TS Technical Specifications and ISO/TR Technical Reports. The presented publications include seven series: ISO 7250, ISO 11226, ISO 11228, ISO 14738, ISO 15534, ISO 15536 and ISO 20685. Most of them are published or in a review stage but they are confirmed as the international standards. Only in a case of such deliverables as: ISO/TR 7250–2:2010, ISO 11228-3:2007, ISO 14738:2002,

ISO 15535:2012 and ISO 20685-2:2015 they need to be revised. The stage of the standard ISO 7250–1:2017 is close of review.

Among these documents the IEEE White Paper can be found which shows findings on the assessment of compatibility and repeatability of traditional anthropometry, the phone apps and body scanners. Moreover, it discusses the application of ISO methods for assessing pairwise compatibility. The learnings will be integrated with the work of the IEEE P3141–3D Body Processing Standards group for future body processing standards (Ballester et al., 2022).

1.5 CONCLUSIONS

Anthropometry dates back to the classical civilizations, but the large-scale development of body measurement started in the 19th century. In the beginning, anthropometry used traditional measuring techniques and tools. However, nowadays depending on the technology, the diverse body scanning systems have become available. They are mainly used to human pattern recognition, special clothing or special human equipment designs. They also allow to document and visualize physical changes, which happen due to training or diseases, and provide exact body sizes for tailor-made solutions.

The use of anthropometric databases requires reliable resources. However, in practice it often appears that they do not only depend on technology but also people collecting these measurements. Therefore, it is advisable to refer to manual, guidelines and standards to know and be able to use the procedures correctly. In spite of the fact that they show voluntary consensus, they are often a part of several laws, codes, and regulations around the globe.

REFERENCES

Anthropometrics Manual - Minnesota Department of Health, 2022, Anthropometrics Manual, https://www.health.state.mn.us/docs/people/wic/localagency/training/nutrition/nst/anthro.pdf.
BALLESTER, A., WRIGHT, W., VALERO, J., SCOTT, E., DEVLIN, T., BULLAS, A., SILVA, J. and MCDONALD, C., 2022, IEEE SA 3D body processing industry connections - comparative analysis of anthropometric methods: Past, present, and future. *Comparative Analysis of Anthropometric Methods: Past, Present, and Future*, 1–52.

BINDAHMAN, S., ZAKARIA, N. and ZAKARIA, N., 2012, 3D Body Scanning Technology: Privacy and Ethical Issues, Proceeding of The International Conference on Cyber Security, Cyber Warfare and Digital Forensic (IEEE CyberSec), pp. 150–154.

BRAGANÇA, S., AREZES, P., CARVALHO, M. and ASHDOWN, S.P., 2016, Current state of the art and enduring issues in anthropometric data collection. *Dyna*, 83(197), 22–30.

BROOKE-WAVELL, K., JONES, P.R.M. and WEST, G.M., 1994, Reliability and repeatability of 3-D body scanner (LASS) measurements compared to anthropometry. *Annals of Human Biology*, 21(6), 571–577.

California Department of Health Care Services, 2016, Systems of Care Division Child Health and Disability Prevention Program, Health Assessment Guidelines, https://www.dhcs.ca.gov/services/chdp/Documents/HAG/4Anthropometric Measure.pdf.

CASADEI, K. and KIEL, J., 2022, Anthropometric measurement. [Updated 2022 September 26]. In StatPearls [Internet] (Treasure Island, FL: StatPearls Publishing). Available from: https://www.ncbi.nlm.nih.gov/books/NBK537315/.

CASHIN, K. and OOT, L., 2018, *Guide To Anthropometry. A Practical Tool for Program Planners, Managers, and Implementers* (Washington, DC: Food and Nutrition Technical Assistance III Project (FANTA)/FHI 360). https://www.fantaproject.org/sites/default/files/resources/FANTA-Anthropometry-Guide-May2018.pdf.

CDC, 2016, National Health and Nutrition Examination Survey (NHANES). Anthropometry Procedure Manal, https://www.cdc.gov/nchs/data/nhanes/nhanes_15_16/2016_anthropometry_ procedures_manual.pdf.

DAANEN, H.A.M. and HAAR, F.B., 2013, 3D whole body scanners revisited. *Displays*, 34(4), 270–275, 2013. https://doi.org/10.1016/j.displa.2013.08.011.

DIANAT, I., MOLENBROEK, J. and CASTELLUCCI, H.I., 2018, A review of the methodology and applications of anthropometry in ergonomics and product design. *Ergonomics*, 12(61), 1–61.

FEATHERS, D.J., PAQUET, V.L. and DRURY, C.G., 2004, Measurement consistency and three-dimensional electromechanical anthropometry. *International Journal of Industrial Ergonomics*, 33, 181–190.

GUPTA, D., 2014, Anthropometry and the design and production of apparel: an overview. In Deepti Gupta, Norsaadah Zakaria (eds.) *Woodhead Publishing Series in Textiles, Anthropometry, Apparel Sizing and Design* (Woodhead Publishing: Cambridge), pp. 34–66.

HAN, H., NAM, Y. and CHOI, K., 2010, Comparative analysis of 3D body scan measurements and manual measurements of size Korea adult females. *International Journal of Industrial Ergonomics*, 40, 530–540.

HELFAND, J., 2019, *Face: A Visual Odyssey* (MIT Press) Cambrigde.

HRDLICKA, A., 1920, Anthropometry, The Wistar Institute of Anatomy and Biology. Available from http://ia700508.us.archive.org/23/items/anthropometry00hrdl/anthropometry00hrdl.pdf. [Accessed on 10 December 2012].

https://www.astm.org/.

https://www.iso.org/.

JONES, P.R.M., WEST, G.M., HARRIS, D.H. and READ, J.B., 1989, The loughborough anthropometric shadow scanner (LASS). *Endeavour*, 13(4), 162–168. https://doi.org/10.1016/S0160–9327(89)80014-6.

KOUCHI, M. and MOCHIMARU, M., 2010, Errors in landmarking and the evaluation of the accuracy of traditional and 3D anthropometry. *Applied Ergonomics*, 1–10. 42(3):518–27.

LAWS, D.R., 2020, Anthropometry: Bertillon's measurement of criminal man. In *A History of the Assessment of Sex Offenders: 1830–2020* (Bingley: Emerald Publishing Limited), pp. 89–97. https://doi.org/10.1108/978-1-78769-359-320201010.

LAU, M.H. and ARMSTRONG, T.J., 2011, The effect of viewing angle on wrist posture estimation from photographic images using novice raters. *Applied Ergonomics*, 42, 634–643.

LIM YC, Abdul Shakor AS and Shaharudin R (2022) Reliability and Accuracy of 2D Photogrammetry: A Comparison With Direct Measurement. Front. Public Health 9:813058. doi: 10.3389/fpubh.2021.813058.

LOVESEY, E.J., 1966, A method for determining facial contours by shadow projection. *Royal Aircraft Establishment Technical Report TR66192*.

MA, L. and NIU, J., 2021, Three-dimensional (3D) anthropometry and its applications in product design. In Karwowski, W. and Salvendy (eds) *Handbook of Human Factors and Ergonomics*, 281–302. https://doi.org/10.1002/9781119636113. Wiley & Sons, Hoboken

MARSHALL, R. and SUMMERSKILL, S., 2019, Posture and anthropometry. In Scataglini, S. and Paul, G. (eds) *DHM and Posturography* (Academic Press), pp. 333–350. London

MINETTO, M.A., PIETROBELLI, A., BUSSO, C., BENNETT, J.P., FERRARIS, A., SHEPHERD, J.A. and HEYMSFIELD, S.B., 2022, Digital anthropometry for body circumference measurements: European phenotypic variations throughout the decades. *Journal of Personalized Medicine*, 12, 906. https://doi.org/10.3390/jpm12060906.

RAJI, R.K., LUO, Q. and LIU, H., 2021, Ergonomics in fashion engineering and design–Pertinent issues. *Work*, 68(1), 87–96.

SAALUDIN, N., SAAD, A. and MASON, C., 2022, Reliability and ethical issues in conducting anthropometric research using 3D scanner technology. In Norsaadah Zakaria (ed) *In The Textile Institute Book Series, Digital Manufacturing Technology for Sustainable Anthropometric Apparel* (Woodhead Publishing) Cambridge, pp. 71–95. https://doi.org/10.1016/B978-0-12-823969-8.00011-3.

SIMS, R.E., MARSHALL, R., GYI, D.E., SUMMERSKILL, S.J. and CASE, K., 2012, Collection of anthropometry from older and physically impaired persons: Traditional methods versus TC2 3-D body scanner. *International Journal of Industrial Ergonomics*, 42(1), 65–72. https://doi.org/10.1016/j.ergon.2011.10.002.

STARK, E., HAFFNER, O. and KU-ERA, E., 2022, Low-cost method for 3D body measurement based on photogrammetry using smartphone. *Electronics*, 11(7), 1048.

The International Fetal and Newborn Growth Consortium, 2012, INTERGROWTH-21st, International Fetal and Newborn Growth Standards for the 21st Century, University of Oxford, https://www.medscinet.net/Interbio/Uploads/ProtocolDocs/Anthropometry%20Handbook.pdf.

TITORIA, S. and SHARMA, D., 2022, Anthropometric profile development for senior Indian women soccer players. *International Journal of Physiology, Nutrition and Physical Education*, 7(2), 147–150.

UNICEF, 2014, Manual for Anthropometry - UNICEF MICS, https://mics.unicef.org

WANG, Z., 2020, Review of real-time three-dimensional shape measurement techniques. *Measurement*, 156, 107624. https://doi.org/10.1016/j.measurement.2020.107624.

WANG, E.M. and CHAO, W., 2010, In searching for constant body ratio benchmarks. *International Journal of Industrial Ergonomics*, 40, 59–67.

WILLIAMS, A.R., 1977, Light Sectioning as a three-dimensional measurement system in medicine. *The Journal of Photographic Science*, 25(2), 85–92. https://doi.org/10.1080/00223638.1977.11737916.

www.en-standard.eu

YU, A., YICK, K.L., NG, S.P. and YIP, J., 2013, 2D and 3D anatomical analyses of hand dimensions for custom-made gloves. *Applied Ergonomics*, 44, 381–392.

ZAKARIA, N. and GUPTA, D. (eds), 2019, *Anthropometry, Apparel Sizing and Design* (London: Woodhead Publishing).

ZIVID, Basic 3D machine vision techniques and principles, https://www.zivid.com/3d-vision-technology-principles.

FURTHER READING

Heymsfield, S. B., Bourgeois, B., Ng, B. K., Sommer, M. J., Li, X., & Shepherd, J. A. (2018). Digital anthropometry: A critical review. *European Journal of Clinical Nutrition*, 72(5), 680–687.

Jenkin, J. (2020). An abridged history of anthropometry. *World Safety Journal, 29*(3), 27–33.

Löffler-Wirth, H., Willscher, E., Ahnert, P., Wirkner, K., Engel, C., Loeffler, M., & Binder, H. (2016). Novel anthropometry based on 3D-bodyscans applied to a large population based cohort. *PloS one, 11*(7), e0159887.

Pomeroy, E., Stock, J. T., & Wells, J. C. (2021). Population history and ecology, in addition to climate, influence human stature and body proportions. *Scientific Reports, 11*(1), 274.

Viviani, C., Arezes, P. M., Braganca, S., Molenbroek, J., Dianat, I., & Castellucci, H. I. (2018). Accuracy, precision and reliability in anthropometric surveys for ergonomics purposes in adult working populations: A literature review. *International Journal of Industrial Ergonomics*, 65, 1–16.

Anthropometry Definition, Uses, and Methods of Measurements

2

R. E. Herron

2.1 DEFINITION

The word "anthropometry" was coined by the French naturalist Georges Cuvier (1769–1832). It was first used by physical anthropologists in their studies of human variability among human races and for comparison of humans to other primates. Anthropometry literally means "measurement of man," or "measurement of humans," from the Greek words ***anthropos***, a man, and ***metron***, a measure.

Although we can measure humans in many different ways, anthropometry focuses on the measurement of bodily features such as body shape and body composition ("static anthropometry"), the body's motion and strength capabilities and use of space ("dynamic anthropometry").

DOI: 10.1201/9781003459767-2

2.2 ORIGINS

The origins of anthropometry can be traced to the earliest humans, who needed information about body parts for many of the same reasons which apply today—to fit clothing, design tools and equipment, etc. No doubt they also used body measurements for other, "nondesign" purposes, such as footprints to estimate the body size of potential adversaries. These and other early applications called for the measurement or estimation of height, as well as the shape and size of hands, feet, and other body parts. Such needs gave rise to the very early use of the terms, "span," "cubit," and "canon," which connote extended arm width, height, and a standard, respectively.

Body proportions were of great interest during classical times, which is clearly evident in the work of artists and sculptors of the period. Around the year 15 BC, the Roman architect Vitruvius wrote about the potential transfer of harmonious body proportions to the design of beautiful buildings. However, it was the work of Renaissance artist anatomists, including Alberti, Pierro della Francesca, Leonardo da Vinci, and especially Albrecht Durer, that ushered in the scientific beginnings of anthropometry. Durer's four-volume publication on human proportions was the first serious attempt to systematize the study of human size and shape.

2.3 USES OF ANTHROPOMETRY

Today, anthropometric measurements are used in a remarkably wide variety of scientific and technical fields, ranging from genetics and nutrition to forensics and industrial design. Within the field of ergonomics, there are myriad applications of anthropometry, primarily associated with different aspects of design for human use.

The goal of ergonomics is to design tools, workplaces, and environments in such a way that humans can function most effectively—in other words, to *optimize* human performance by achieving the best possible fit between the human operator, the equipment (hardware and software), and the working environment (physical and psychosocial). This fit is often referred to as "the human–machine interface." Anthropometry can and does play a major role in achieving this goal because variations in bodily features, such as shape, size, strength, and reach, affect the way people perform tasks and, thus, have an important influence on whether the human–machine interface is a good one.

The breadth of possible applications of anthropometry for improving the human interface is remarkably wide-ranging, from industrial equipment, clothing, and furniture, to surgical tools, farm implements, aircraft controls, and virtually every item in the environment with which humans interact.

Over the years, engineers, designers, architects, and others who design products or processes have increasingly recognized the need for anthropometric data on the users of their creations. Of course, the need for anthropometric information and the type of data required varies greatly from one application to the other. In some areas, the fit is "soft," as in a loose garment such as a bathrobe; in other areas, the fit is "hard," e.g. in a respirator for protection against breathing toxic fumes. The fit of the bathrobe can be an approximation and still serve its intended purpose, whereas the respirator mask must conform closely to the geometry of the face in order to maintain adequate contact and prevent leakage. In the case of the bathrobe, data on height and a few body girth measurements for the prospective user population may be all the information needed to ensure adequate body coverage for a good interface. However, for the respirator mask, it may be necessary to obtain detailed three-dimensional measurements of individual facial geometry to ensure a satisfactory fit.

2.4 METHODS

2.4.1 Traditional Methods

The earliest methods and tools for making anthropometric measurements were very simple, but they were quite effective and some of these rudimentary devices, such as measuring sticks and calipers, have endured to the present day. The types of data that result from the simple tools shown in Figure 2.1 are quite varied and include numerous length measurements—e.g. height, various widths, such as the shoulder and pelvis, and circumferences, such as the waist and chest.

These tools appear to be deceptively easy to use but, in fact, when used for scientific or engineering purposes, they require a high level of care in order to achieve acceptable levels of validity and reliability. For example, it is important to know where to locate and how to align the measuring device (tape, caliper, measuring stick or whatever) on the body surface and to do so in a consistent, reproducible manner. In using a measuring tape, the tension in the tape and the degree of tissue compression must be suitably controlled.

Over the last hundred years, many volumes have been written about how to perform traditional anthropometric measurements and the reader is referred to the list of publications given below for further details.

2.4.2 Modern Methods

Over the last 30 years the tools of anthropometry have changed dramatically, propelled by advances in computer and shape-sensing technology. Traditional linear measures of the body and body parts have gradually given way to 3-D and 4-D measures, computer models and, most recently, fractals, using new multi-dimensional sensing devices which capture considerably more of the subtle variations of human form and function.

The transformation of traditional anthropometry took a major turn in the 1960s with the growing recognition that the human body is an irregular, three-dimensional, dynamic organism, which calls for different strategies of mathematical abstraction and different measuring instruments than those used for obtaining traditional linear dimensions. This development led to greater use of mapping approaches, using contours and coordinates, and mathematical approaches based on polynomials, nurbs, b-splines, and other strategies for representing the irregular, multi-dimensional bodily features. A recent conference report (Vannier *et al.* 1992) provides a valuable overview of these 3-D anthropometric methods and their future prospects.

Traditional anthropometry has been largely confined to surface measurements of the body, with the exception of a rather limited use of X-rays (Figure 2.2). The recent rapid growth of Computed Tomography (CT), Magnetic Resonance Imaging (MRI), Positron Emission Tomography (PET), and other new medical imaging devices has exposed the potential for generating

FIGURE 2.1   An array of traditional anthropometric measurement tools.

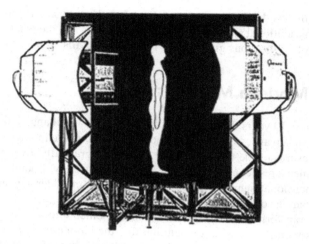

FIGURE 2.2   Modern anthropometric measurement methods.

previously inaccessible anthropometric data on internal body structures and functions. Obviously, knowing the dimensions and the motions of internal body parts can provide invaluable information for many ergonomic needs. For example, it would certainly be helpful to know what happens to the geometry of internal organs and systems during the performance of various tasks and when the body adopts different postures.

Other recent developments in anthropometric recording and instrumentation strategies include the use of 3-D images for matching body parts in reconstructive surgery, automated hip surgery, measurement of heart and

2.5 ANTHROPOMETRY: DEFINITION, USES, AND METHODS OF MEASUREMENT

other organ volumes for transplants, stereometric brain surgery, and dose radiation measurements. Although, it may not be immediately obvious, all of these recent applications can translate into new potentials for advancing the use of anthropometry for ergonomic design and fitting purposes.

The use of 3-D and 4-D computer models to represent human form and function is still in an early stage, but already over 200 models have been developed for various ergonomic applications. Interest in creating more accurate and more versatile computer models of the human body has received a major impetus from the growing use of animation in the motion picture industry.

This line of activity can be expected to grow dramatically in the near future as better (more comprehensive and statistically valid) multi-dimensional body data become available and the power of micro-computers continues to grow. Having a wide range of human models covering myriad details of human form and function stored in an immediately accessible form would seem to be the preferred *modus operandi* for helping engineers and designers to extend their uses of anthropometry for ergonomic purposes.

2.6 WORK SPACE-ENVELOPE MEASUREMENT

The measurement of the work space-envelope for different occupational activities is an important aspect of ergonomic design. The space-envelope occupied by the body while performing a task is larger than the space taken up by the body itself. The first measurements of this type were made using simple mechanical devices but, over the last 20 years, this approach has been superceded by non-contact, three-dimensional video imaging techniques. The video methods are portable, inexpensive, and easy to use.

2.7 STRENGTH MEASUREMENT

The measurement of human strength is important for the design of tools and equipment and other ergonomic applications. The growing availability of modern electronic dynamometers and strain gauges linked to computers has made the taking of human strength measurements a relatively simple operation; however, the acquisition of valid and reliable strength data requires considerable care and skill. Details about the different methods for measuring human strength can be found in the publications listed below.

2.8 RANGES OF ANTHROPOMETRIC DATA

In many ergonomic applications, it is necessary to know the ranges of pertinent anthropometric measurements found among a particular population, e.g. the prospective users of a new piece of equipment. The words "average" and "percentile" often appear in discussions about such matters, and the way these terms are used in anthropometry requires special attention. Although a single bodily characteristic, such as height or weight, can be expressed as average or at a particular percentile level, e.g. at the 50th percentile level, there is no such thing as an "average" human or a "50th percentile human." This stems from the fact that the position of each individual relative to an average value or a percentile scale varies from one bodily feature to the other. An individual's height may be at the 75th percentile level (i.e. 75% of the population are the same or of lesser height), but his weight may be at the 60th percentile and his chest girth at the 50th percentile. Thus, there is no such thing as a 95th percentile human—except as it relates to a specific bodily feature—because combining several 95th (or any other single) percentile values for various dimensions on the same body produces an unrealistic human form. Anthropometric variables can be combined, but this requires the use of multivariate and other statistical methods, which are described in one or more of the publications listed below. These references also discuss the correct use of means, medians, standard deviations, standard errors, and other statistics for ergonomic purposes.

Another major issue in obtaining or producing representative anthropometric data relates to problems of population sampling. For example, if the goal is to design a piece of equipment for the general US population and there is a need to know the range of body shapes and sizes, there is no extant database of such information. The available anthropometric databases do not include a statistically valid sample of the US population. Therefore, it is often necessary to find a compromise, which might involve combining anthropometric information from various sources, such as military populations, and from limited samples of the general population. Furthermore, the ergonomic needs of special populations such as the elderly, disabled, infant/child, and ethnic groups often call for anthropometric data which are representative of the particular segment or segments of the population.

Sampling strategies and tactics are explained in more detail in the references given below. However, the current problems in this area will not be alleviated in any significant way until a comprehensive anthropometric survey of the US population is completed. Recently, several countries have conducted national anthropometric surveys and others plan to follow suit in the near future. The cost and logistical difficulties involved in such surveys will probably limit them to the more prosperous nations for the foreseeable future, so that the development of a truly global anthropometric data base is unlikely for many years to come.

2.9 THE FUTURE OF ANTHROPOMETRY

Although the tools of modern anthropometry have already become quite sophisticated, no doubt the future will bring even more dramatic "high-tech" advances. These will include a wide range of new sensors (non-contact, multi-dimensional sensors, etc.) to measure the changes (static and dynamic, as well as internal and external) in body shape and size which accompany everyday activities, as well as those which are associated with growth, disease, aging, etc. Computer body models will reach a new level of realism based on more comprehensive and valid population data on a wider range of useful measures.

Another noteworthy development that illustrates what the future holds is the National Library of Medicine project launched in 1992 to create 3-D "visible humans." Each visible human is graphically reconstructed from the images of a series of fine cross-sections, taken from head to foot, of two "representative" human cadavers—one male and one female. The first dissections of this type were conducted by Eycleshymer and Schoemaker in 1911 when they selected a "representative" series of cross-sections from a sample of 50 individuals. In 1974, a group at the Biostereometrics Laboratory, Baylor College of Medicine, similarly sliced a cadaver into 92 cross-sections, using a specially designed saw. The images of successive cross-sections were then used to reconstruct the 3-D geometry of selected internal organs and systems for use by army ballistics researchers studying what body tissues would be affected by different missiles entering the body at various locations and angles. Already, the National Library of Medicine visible humans have been widely used and have met important needs for teaching and other purposes, but their representativeness of anthropometric variation among humans is obviously quite limited.

Future anthropometric methods will cope better with the fact that humans are not fixed objects like statues, but are dynamic organisms whose structures and postures change daily and throughout life—these methods will recognize that there are no fixed points on the body, just ever-changing irregular forms. We must exploit the potential of more "holistic" parsimonious mathematical abstractions for representing, measuring, analyzing, and interpreting the subtleties of human form and function, both internally and externally, from conception to death. Comprehensive multi-dimensional human models based on statistically valid anthropometric data for a wide variety of populations and specialized groups will be instantly available for use by engineers, designers, and others, in their computers and virtual-reality simulators. Such a development will help to elevate the science and technology of anthropometry to a new level of precision and utility.

FURTHER READING

Bartol, K., Bojanić, D., Petković, T., & Pribanić, T. (2021). A review of body measurement using 3D scanning. *Ieee Access*, *9*, 67281–67301.

Kaashki, N. N., Hu, P., & Munteanu, A. (2021). Anet: A deep neural network for automatic 3d anthropometric measurement extraction. *IEEE Transactions on Multimedia*, *25*, 831–844, 2023. Doi: 10.1109/TMM.2021.3132487.

Loeffler-Wirth, H., Vogel, M., Kirsten, T., Glock, F., Poulain, T., Körner, A., Loeffler, M. Kiess, W. & Binder, H. (2017). Body typing of children and adolescents using 3D-body scanning. *PLoS One*, *12*(10), e0186881.

Nariño Lescay, R., Alonso Becerra, A., & Hernández González, A. (2016). Anthropometry. Comparative analysis of technologies for the capture of anthropometric dimensions. *Revista EIA*, *26* (July/December), 47–59.

Rumbo-Rodríguez, L., Sánchez-SanSegundo, M., Ferrer-Cascales, R., García-D'Urso, N., Hurtado-Sánchez, J. A., & Zaragoza-Martí, A. (2021). Comparison of body scanner and manual anthropometric measurements of body shape: A systematic review. *International Journal of Environmental Research and Public Health*, *18*(12), 6213.

Keyword
Body Sizes of Americans

3

K. H. E. Kroemer

The most recent complete set of data describing the body sizes of American adults derives from the survey of US Army soldiers by Gordon *et al.* (1989). They followed the traditional measurement descriptions and procedures (Roebuck 1995). Table 3.1 lists 37 of their measurements, which are of particular interest to the designer for their applications to human–machine systems, equipment, and tools. Since the anthropometric data stem from subjects in standardized postures (body angles at 0, 90, or 180 degrees), many numbers need conversions to

The body sizes of US soldiers reported by Gordon *et al.* (1989) are still the best possible estimates of American civilian adults because most of the results of a newer study by Robinette *et al.* (2002) are proprietary and not available to the public. In the US military, the Army is the least selective service and hence best presents the general population. Table 3.2 contains the data measured on the soldiers. In the absence of better information, their anthropometry may be taken as representative of the general population of American civilian adults. Yet, caution is in order when applying the weight data, listed as no. 37 (and related dimensions such as no. 23, abdominal depth, or 28, hip breadth): given the current trend toward obesity, the actual mean and standard deviation are likely to be larger than shown in Table 3.2.

DOI: 10.1201/9781003459767-3

TABLE 3.1 Anthropometric Measures and their Applications

DIMENSIONS	APPLICATIONS
Stature	
The vertical distance from the floor to the top of the head, when standing. [99]	A main measure for comparing population samples. Reference for the minimal height of overhead obstructions Add height for more clearance, hat, shoes, stride.
Eye Height, Standing	
The vertical distance from the floor to the outer corner of the right eye, when standing. [D19]	Origin of the visual field of a standing person. Reference for the location of visual obstructions and of targets such as displays; consider slump and motion.
Shoulder Height (Acromion), Standing	
The vertical distance from the floor to the tip (acromion) of the shoulder, when standing. [2]	Starting point for arm length measurements; near the center of rotation of the upper arm. Reference point for hand reaches; consider slump and motion.
Elbow Height, Standing	
The vertical distance from the floor to the lowest point of the right standing, with the elbow flexed at 90 degrees. [D16]	Reference for height and distance of the work area of the hand and elbow, when the location of controls and fixtures; consider slump and motion.
Hip Height (Trochanter), Standing	
The vertical distance from the floor to the trochanter landmark on of the right thigh, when standing. [107]	Traditional anthropometric measure, indicator of leg length and the upper side height of the hip joint. Used for comparing population samples.
Knuckle Height, Standing	
The vertical distance from the floor to the knuckle (metacarpal bone) of the middle finger of the right hand, when standing.	Reference for low locations of controls, handles, and handrails; consider slump and motion of the standing person.
Fingertip Height, Standing	
The vertical distance from the floor to the tip of the extended index right hand, when standing. [D13]	Reference for the lowest location of controls, handles, and handrails; finger of the consider slump and motion of the standing person.
Sitting Height	
The vertical distance from the sitting surface to the top of the head, when sitting. [93]	Reference for the minimal height of overhead obstructions. Add height when for more clearance, hat, trunk motion of the seated person.

(Continued)

TABLE 3.1   (Continued)

DIMENSIONS	APPLICATIONS
Sitting Eye Height	
The vertical distance from the sitting surface to the outer corner of the right eye, when sitting. [49]	Origin of the visual field of a seated person. Reference point for the location of visual obstructions and of targets such as displays; consider slump and motion.
Sitting Shoulder Height (Acromion)	
The vertical distance from the sitting surface to the tip (acromion) of the shoulder, when sitting. [3]	Starting point for arm length measurements; near the center of rotation of the upper arm. Reference for hand reaches; consider slump and motion.
Sitting Elbow Height	
The vertical distance from the sitting surface to the lowest point of the right elbow, when sitting, with the elbow flexed at 90 degrees. [48]	Reference for the height of an armrest, of the work area of the hand and of keyboard and controls; consider slump and motion of the seated person.
Sitting Thigh Height (Clearance)	
The vertical distance from the sitting surface to the highest point on the horizontal right thigh, with the knee flexed at 90 degrees. [104]	Reference for the minimal clearance needed between seat pan and the top of underside of a structure, such as a table or desk; add clearance for clothing and motions.
Sitting Knee Height	
The vertical distance from the floor to the top of the right kneecap, sitting, with the knees flexed at 90 degrees. [73]	Traditional anthropometric measure for lower leg length. Reference for when the minimal clearance needed below the underside of a structure, such as a table or desk; add height for shoe.
Sitting Popliteal Height	
The vertical distance from the floor to the underside of the thigh directly behind the right knee; when sitting, with the knees flexed at 90 degrees. [86]	Reference for the height of a seat; add height for shoe.
Shoulder–Elbow Length	
The vertical distance from the underside of the right elbow to the right acromion, with the elbow flexed at 90 degrees and the upper arm hanging vertically. [91]	Traditional anthropometric measure for comparing population samples.

(Continued)

TABLE 3.1   (Continued)

DIMENSIONS	APPLICATIONS

Elbow–Fingertip Length

The distance from the back of the right elbow to the tip of the extended middle finger, with the elbow flexed at 90 degrees. [54]

Traditional anthropometric measure. Reference for fingertip reach when moving the forearm in the elbow.

Overhead Grip Reach, Sitting

The vertical distance from the sitting surface to the center of a cylindrical rod firmly held in the palm of the right hand. [D45]

Reference for the height of overhead controls operated by a seated person. Consider ease of motion, reach, and finger/hand/arm strength.

Overhead Grip Reach, Standing

The vertical distance from the standing surface to the center of a cylindrical rod firmly held in the palm of the right hand. [D42]

Reference for the height of overhead controls operated by a standing person. Add shoe height. Consider ease of motion, reach, and finger/hand/arm strength.

Forward Grip Reach

The horizontal distance from the back of the right shoulder blade to the center of a cylindrical rod firmly held in the palm of the right hand. [D21]

Reference for forward reach distance. Consider ease of motion, reach, and finger/hand/arm strength.

Arm Length, Vertical

The vertical distance from the tip of the right middle finger to the right acromion, with the arm hanging vertically. [D3]

A traditional measure for comparing population samples. Reference for the location of controls very low on the side of the operator. Consider ease of motion, reach, and finger/hand/arm strength.

Downward Grip Reach

The vertical distance from the right acromion to the center of a cylindrical rod firmly held in the palm of the right hand, with the arm hanging vertically. [D43]

Reference for the location of controls low on the side of the operator. Consider ease of motion, reach, and finger/hand/arm strength.

Chest Depth

The horizontal distance from the back to the right nipple. [36]

A traditional measure for comparing population samples. Reference for the clearance between seat backrest and the location of obstructions in front of the trunk.

(Continued)

TABLE 3.1   (Continued)

DIMENSIONS	APPLICATIONS

Abdominal Depth, Sitting

The horizontal distance from the back to the most protruding point on the abdomen [1]

A traditional measure for comparing population samples. Reference for the clearance between seat backrest and the location of obstructions in front of the trunk.

Buttock–Knee Depth, Sitting

The horizontal distance from the back of the buttocks to the most protruding point on the right knee, when sitting with the knees flexed at 90 degrees. [26]

Reference for the clearance between seat backrest and the location of obstructions in front of the knees.

Buttock–Popliteal Depth, Sitting

The horizontal distance from the back of the buttocks to back of the right knee just below the thigh, when sitting with the knees flexed at 90 degrees. [27]

Reference for the depth of a seat.

Shoulder Breadth (Biacromial)

The distance between the right and left acromion. [10]

A traditional measure for comparing population samples. Indicator of the distance between the centers of rotation of the two upper arms.

Shoulder Breadth (Bideltoid)

The maximal horizontal breadth across the shoulders between the lateral margins of the right and left deltoid muscles. [12]

Reference for the lateral clearance required at shoulder level. Add space for ease of motion and tool use.

Hip Breadth, Sitting

The maximal horizontal breadth across the hips or thighs, whatever is greater, when sitting. [66]

Reference for seat width. Add space for clothing and ease of motion.

Span

The distance between the tips of the middle fingers of the horizontally outstretched arms and hands. [98]

A traditional measure for comparing population samples. Reference for sideway reach.

Elbow Span

The distance between the tips of the elbows of the horizontally outstretched upper arms when the elbows are flexed so that the fingertips of the hands meet in front of the trunk.

Reference for the lateral space needed at upper body level for ease of motion and tool use.

(Continued)

TABLE 3.1 (Continued)

DIMENSIONS	APPLICATIONS
Head Length	
The distance from the glabella (between the browridges) to the most rearward protrusion (the occiput) on the back, in the middle of the skull. [62]	A traditional measure for comparing population samples. Reference for head gear size.
Head Breadth	
The maximal horizontal breadth of the head above the attachment of the ears. [60]	A traditional measure for comparing population samples. Reference for head gear size. A traditional measure for comparing population samples. Reference for hand tool and gear size. Consider manipulations, gloves, tool use.
Hand Length	
The length of the right hand between the crease of the wrist and the tip of the middle finger, with the hand flat. [59]	A traditional measure for comparing population samples. Reference for hand tool and gear size, and for the opening through which a hand may fit. Consider manipulations, gloves, tool use.
Hand Breadth	
The breadth of the right hand across the knuckles of the four fingers. [57]	A traditional measure for comparing population samples. Reference for hand tool and gear size, and for the opening through which a hand may fit. Consider manipulations, gloves, tool use.
Foot Length	
The maximal length of the right foot, when standing. [51]	A traditional measure for comparing population samples. Reference for shoe and pedal size.
Foot Breadth	
The maximal breadth of the right foot, at right angle to the long axis the foot, when standing. [50]	A traditional measure for comparing population samples. Reference of for shoe size, spacing of pedals.
Weight (in kg)	
Nude body weight taken to the nearest tenth of a kilogram.	A traditional measure for comparing population samples. Reference for body size, clothing, strength, health, etc. Add weight for clothing and equipment worn on the body.

Descriptions of the measures according to Gordon, Churchill, Clauser et al. (1989) with their numbers in brackets.

TABLE 3.2 Anthropometry of US Adults (in mm)

DIMENSION	MEN				WOMEN			
	5TH PERCENTILE	MEAN	95TH PERCENTILE	SD	5TH PERCENTILE	MEAN	95TH PERCENTILE	SD
Stature [99]	1647	1756	1867	67	1528	1629	1737	64
Eye height, standing [D19]	1528	1634	1743	66	1415	1516	1621	63
Shoulder height (acromion), standing [2]	1342	1443	1546	62	1241	1334	1432	58
Elbow height, standing [D16]	995	1073	1153	48	926	998	1074	45
Hip height (trochanter) [107]	853	928	1009	48	789	862	938	45
Knuckle height, standing	Na	na	na	na	na	na	na	na
Fingertip height, standing [D13]	591	653	716	40	551	610	670	36
Sitting height [93]	855	914	972	36	795	852	910	35
Sitting eye height [49]	735	792	848	34	685	739	794	33
Sitting shoulder height (acromion) [3]	549	598	646	30	509	556	604	29
Sitting elbow height [48]	184	231	274	27	176	221	264	27

(Continued)

TABLE 3.2 (Continued)

DIMENSION	MEN						WOMEN					
	5TH PERCENTILE	MEAN	95TH PERCENTILE	SD	5TH PERCENTILE	MEAN	95TH PERCENTILE	SD
Sitting thigh height (clearance) [104]	149	168	190	13	140	160	180	12
Sitting knee height [73]	514	559	606	28	474	515	560	26
Sitting popliteal height [86]	395	434	476	25	351	389	429	24
Shoulder-elbow length [91]	340	369	399	18	308	336	365	17
Elbow-fingertip length [54]	448	484	524	23	406	443	483	23
Overhead grip reach, sitting [D45]	1221	1310	1401	55	1127	1212	1296	51
Overhead grip reach, standing [D42]	1958	2107	2260	92	1808	1947	2094	87
Forward grip reach [D21]	693	751	813	37	632	686	744	34
Arm length, vertical [D3]	729	790	856	39	662	724	788	38
Downward grip reach [D43]	612	666	722	33	557	700	664	33
Chest depth [36]	210	243	280	22	209	239	279	21
Abdominal depth, sitting [1]	199	236	291	28	185	219	271	26
Buttock-knee depth, sitting [26]	569	616	667	30	542	589	640	30

Measurement								
Buttock-popliteal depth, sitting [27]	458	500	546	27	440	482	528	27
Shoulder breadth (biacromial) [10]	367	397	426	18	333	363	391	17
Shoulder breadth (bideltoid) [12]	450	492	535	26	397	433	472	23
Hip breadth, sitting [66]	329	367	412	25	343	385	432	27
Span [98]	1693	1823	1960	82	1542	1672	1809	81
Elbow span	Na	na	na	na	na	na	na	na
Head length [62]	185	197	209	7	176	187	198	6
Head breadth [60]	143	152	161	5	137	144	153	5
Hand length [59]	179	194	211	10	165	181	197	10
Hand breadth [57]	84	90	98	4	73	79	86	4
Foot length [51]	249	270	292	13	224	244	265	12
Foot breadth [50]	92	101	110	5	82	90	98	5
Weight (kg)	62	79	98	11	50	62	77	8

Measurements were taken in 1987/8 on US Army soldiers, 1774 men and 2208 women, by Gordon, Churchill, Clauser et al. (1989), who used the numbers in brackets.

REFERENCES

GORDON, C.C., CHURCHILL, T., CLAUSER, C.E., BRADTMILLER, B., MCCONVILLE, J.T., TEBBETTS, I. and WALKER, R.A., 1989, *1988 Anthropometric Survey of US Army Personnel. Summary Statistics Interim Report.* Technical Report NATICK/TR-89–027 (Natick, MA: United States Army Natick Research, Development and Engineering Center).

ROBINETTE, K.M., BLACKWELL, S., DAANEN, H., BOEHMER, M., FLEMING, S., BRILL, T., HOEFERLIN, D. and BURNSIDES, D. 2002, *Civilian American and European Surface Anthropometry Resource (CAESAR) Final Report*, Volume I: Summary (AFRL-HE-WP-TR-2002–0169) (Wright-Patterson AFB, OH: United States Air Force Research Laboratory).

ROEBUCK, J.A., 1995, *Anthropometric Methods* (Santa Monica, CA: Human Factors and Ergonomics Society).

FURTHER READING

Cámara, A. D., Martínez-Carrión, J. M., Puche, J., & Ramon-Muñoz, J. M. (2019). Height and inequality in Spain: A long-term perspective. *Revista de Historia Económica-Journal of Iberian and Latin American Economic History, 37*(2), 205–238.

Choong, C. L., Alias, A., Abas, R., Wu, Y. S., Shin, J. Y., Gan, Q. F., Thu, K.M., & Choy, K. W. (2020). Application of anthropometric measurements analysis for stature in human vertebral column: A systematic review. *Forensic Imaging, 20*, 200360.

El Kari, K., Mankai, A., Kouki, D. E., Mehdad, S., Benjeddou, K., El Hsaini, H., El Mzibri, M., & Aguenaou, H. (2023). Anthropometry-based prediction equation of body composition in a population aged 12–88 Years. *The Journal of Nutrition.* 153(3), 657–664.

Kuebler, T., Luebke, A., Campbell, J., & Guenzel, T. (2019). Size North America–The New North American anthropometric survey. In *Digital Human Modeling and Applications in Health, Safety, Ergonomics and Risk Management. Human Body and Motion: 10th International Conference, DHM 2019, Held as Part of the 21st HCI International Conference, HCII 2019, Orlando, FL, USA, July 26–31, 2019, Proceedings, Part I 21* (pp. 88–98). Springer International Publishing.

Xia, S., & Istook, C. (2017). A method to create body sizing systems. *Clothing and Textiles Research Journal, 35*(4), 235–248.

Anthropometric Databases

4

R. E. Herron

4.1 INTRODUCTION

Anthropometry literally means the measurement of humans. Although one can measure humans in many different ways, anthropometry focuses on the measurement of bodily features such as shape and size ("static anthropometry"), body motion, use of space and physical capacities such as strength ("functional anthropometry"). A database is simply a compilation of information, generally used for reference purposes.

There are several types of database in ergonomics. They include: (1) much of the published literature itself, (2) bibliographic aids containing citations, abstracts and indexes to the literature and (3) manuals, guides, handbooks and other compilations that summarize the basic data and methods of the field. Anthropometric databases include all of these formats, but the most widely used are in the form of manuals, guides, handbooks and statistical compilations that attempt to summarize the variability in anthropometric characteristics of different human populations or groups. A new type of anthropometric database—the computer-generated human model—is a rapidly growing phenomenon that is leading to major changes in how anthropometric information is used for ergonomic design and other purposes.

DOI: 10.1201/9781003459767-4

4.2 HISTORY OF ANTHROPOMETRIC DATABASES

Interest in applying anthropometric data to the design of tools, instruments, equipment and systems has a long history. Early human societies employed folk norms for the design of digging, cutting and shaping tools. Natural units of measurement first developed around the hand, fingers and foot and were not replaced with physical units of measurement until many years later. The systematic collection of anthropometric data on large groups of individuals did not occur until the 18th century with the work of Linne and Buffon. The industrial revolution increased the demand for pertinent body dimensions to meet the needs of such mass markets as furniture and clothing. In 1883 Sir Francis Galton undertook a detailed, quantitative survey of anthropometric and other characteristics of over 9,000 visitors to the London Health Exposition. However, it was not until World War II, when the human–technology interface reached new levels of complexity, that the need for better anthropometric data for workstation and equipment design became a major focus of attention. Today, there is no longer any doubt that the wide and constantly growing variety of ergonomic design problems calls for more comprehensive, versatile, parsimonious, valid and reliable anthropometric data sets for use in representing the shapes, sizes and functional capacities of target user populations.

4.3 HOW ARE ANTHROPOMETRIC DATABASES USED IN ERGONOMICS?

In the majority of real-world ergonomic design applications involving the interaction of humans with tools, instruments and systems, there is a need for information about the population of users. Very often the need is for anthropometric data. Differences in body shape, size and physical capacities are among the most obvious manifestations of human variability. It is clearly apparent that people differ in height, weight and arm length, as well as in the other myriad ways in which body form and function can be measured. A well-designed or selected anthropometric database can provide vitally important information for optimizing the human–machine–equipment or system interface.

Creating or selecting and using an anthropometric database effectively is not a trivial matter. There are many hazards, which can undermine the best intentions of the unwary practitioner. First, one has to define which anthropometric parameters are relevant and necessary. Determining which bodily dimensions are important starts with a clear understanding of the task being performed. Then, the user population must be identified and characterized, in terms of homogeneity (e.g. highly selected astronauts) or the lack of it—heterogeneity (e.g. more diverse assembly line workers), as well as other relevant demographic variables, such as age and sex. An initial search should be made to locate a database that matches the target user population as closely as possible. Unfortunately, although hundreds of anthropometric databases are now available, the prospects of finding a good match (from a valid and reliable source) are not good—due to the wide diversity of today's working population, in terms of racial, ethnic, sex, age, nutritional and other variables. This ever-changing diversity is the source of numerous human factors problems, due to the difficulty in defining exactly what the makeup of the civilian population really is. Wide-ranging differences in body shape and size of various ethnic groups are of special significance in the design and manufacturing of products for the global marketplace.

But the current utility of anthropometric databases is not all doom and gloom. Notwithstanding the low probability of finding a close match between existing databases and various target populations, informed and useful compromises can be made. This involves supplementing or adjusting the best available data set, using sophisticated methods for estimating or forecasting, to arrive at a description of the desired design criteria—the range of relevant dimensional measurements which would be expected in the target population. While these strategies can be productive in generating usable and useful anthropometric data for design purposes, a well-conceived and executed methodology is essential. Further details about such matters can be found in the references.

4.4 WHAT ANTHROPOMETRIC DATABASES ARE AVAILABLE?

There is a vast supply of anthropometric databases on military populations, but unfortunately those on civilian populations are fewer. The main sources of military data are publications of government agencies and specialized centers such as the Air Force, Army, NASA and CSERIAC (Crew Systems Ergonomics Information Analysis Center). Many such publications can be obtained from the National Technical Information Service (NTIS) or the Defense Technical Information Center.

The scarcity of good anthropometric databases on the US civilian population is due to a number of limiting factors: (1) inadequate measurement technologies, (2) costs and (3) logistical problems—such as difficulty in mobilizing and accessing representative populations. Fortunately, the situation is slowly improving. In the meantime, one can locate the best extant anthropometric databases via the references.

4.5 WHAT NEW ANTHROPOMETRIC DATABASES ARE UNDER DEVELOPMENT?

Better civilian anthropometric databases, which are more representative of today's diverse population, are on the horizon, largely due to a growing demand for anthropometric data and the continuing evolution of more practical computer-based anthropometric measurement instruments.

These new compilations will go far beyond traditional linear dimensional (2-D) data and include the much more comprehensive and informative multidimensional (3-D, 4-D, etc.), biostereometric data.

The use of a biostereometric approach permits the storing of digital geometric descriptions of entire body forms, i.e. 3-D spatial replicas). These archival databases can be mined for an infinite variety of anthropometric data to meet a wide range of user needs. Another advantage is the ease with which such computer-based compilations can be maintained and up-dated to ensure continuing representative fidelity of the total population and various subpopulations.

4.6 COMPUTER MODELS OF HUMAN FUNCTIONAL CAPACITIES

Another consequence of using a biostereometric approach to the measurement of body geometry is the development of more realistic, animated human models. Such multidimensional models constitute another type of anthropometric database which is changing how anthropometric data are used for ergonomic design applications. However, the biostereometric data on which these models are based do not yet adequately represent the variations in body geometry and motion patterns which are present in the civilian or military populations. This is largely due to the continuing inadequacies of 3-D scanning methods for gathering the basic body geometry and motion data. As these methods become more cost-effective, portable, faster, user-friendly and technologically sophisticated, the current deficiencies will be overcome.

Ultimately, highly representative computer models can be installed in computer-aided design (CAD) platforms and accessed instantaneously to serve a wide variety of ergonomic design functions. The growing availability of such "virtual humans" can overcome many of the problems associated with current anthropometric databases and help promote the proactive use of anthropometric information for improving the human–machine and human–environment design interface.

4.7 FURTHER READING

There is an extensive literature related to the subject of anthropometric databases. The references constitute but a small sample of the extant sources. However, they should at least provide a point of entry for obtaining further information about an area of ergonomics which is entering its most exciting phase—when anthropometric databases of realistic, multidimensional, human models are found in the computers of every designer who is concerned with ergonomic design for human use.

Cheng, I. F., Kuo, L. C., Lin, C. J., Chieh, H. F., & Su, F. C. (2019). Anthropometric database of the preschool children from 2 to 6 Years in Taiwan. *Journal of Medical and Biological Engineering*, *39*, 552–568.

Dawal, S. Z. M., Ismail, Z., Yusuf, K., Abdul-Rashid, S. H., Shalahim, N. S. M., Abdullah, N. S., & Kamil, N. S. M. (2015). Determination of the significant anthropometry dimensions for user-friendly designs of domestic furniture and appliances–Experience from a study in Malaysia. *Measurement*, *59*, 205–215.

ISO 15535. General requirements for establishing anthropometric databases

ISO 20685 (all parts), 3D scanning methodologies for internationally compatible anthropometric databases

Vinué, G. (2017). Anthropometry: An R package for analysis of anthropometric data. *Journal of Statistical Software*, *77*, 1–39.

Engineering Anthropometry

5

K. H. E. Kroemer

People come in a variety of sizes, and their bodies are not assembled in the same proportions. Thus, fitting clothing, equipment, or workstations to suit the body requires careful consideration; design for the statistical "average" will not do. Instead, for each body segment to be fitted, the designer must determine the critical dimension(s). Design for fitting clothing, tools, workstations, and equipment to the body usually follows these steps:

Step 1: Select those anthropometric measures that directly relate to defined design dimensions. Examples are: hand length related to handle size; shoulder and hip breadth related to escape-hatch diameter; head length and breadth related to helmet size; eye height related to the heights of windows and displays; knee height and hip breadth related to the legroom in a console.

Step 2: For each of these pairings, determine whether the design must fit only one given percentile (minimal or maximal) of the body dimension, or must fit a range along that body dimension. Examples are: the escape hatch must be big enough to accommodate the largest extreme value of shoulder breadth and hip breadth, considering clothing and equipment worn; the handle size of pliers is probably selected to fit a smallish hand; the legroom of a console must accommodate the tallest knee heights; the height of a seat should be adjustable to fit persons with short and with long lower legs. (See below how to determine percentiles.)

Step 3: Combine all selected design values in a careful drawing, mock-up, or computer model to ascertain that they are compatible.

DOI: 10.1201/9781003459767-5

For example, the required legroom clearance height needed for sitting persons with long lower legs may be very close to the height of the working surface determined from elbow height.

Step 4: Determine whether one design will fit all users. If not, several sizes or adjustment must be provided to fit all users. Examples are: one extra-large bed size fits all sleepers; gloves, and shoes must come in different sizes; seat heights of office chairs are adjustable.

Percentiles can be determined either by estimation or by calculation.

Estimation is appropriate when the data set is not normally distributed or too small. In this case, the data point may be estimated by counting, weighing, or sampling measurement according to best possible judgment.

Calculation uses statistical considerations.

Just two simple statistics fully describe a normally distributed set of n data (a Gaussian distribution): the first descriptor is the mean m (also commonly called average):

$$m = \frac{\sum x}{n}$$

where $\sum x$ is the sum of the individual measurements. (In a normal distribution, mean, mode, median, 50th percentile, all coincide.)

The second descriptor is the standard deviation (SD), which describes the distribution of the data:

$$SD = \left[\frac{\sum (x-m)^2}{n-1} \right]^{\frac{1}{2}}$$

It is often useful to describe the variability of a sample by dividing the SD by the mean m. The resulting coefficient of variation (CV) (in percent) is:

$$CV = \frac{100 \; SD}{m}$$

To calculate a percentile value p of a normal distribution you simply multiply the SD by a factor k, selected from Table 5.1. Then, if p is below the mean, subtract the product from the mean m:

$$p = m - k \times SD$$

TABLE 5.1 Factor *k* for Computing Percentiles from Mean *m* and Standard Deviation SD

K	*PERCENTILE p LOCATED ABOVE THE MEAN m* $p = m - k \times SD$	*PERCENTILE p LOCATED ABOVE THE MEAN m* $p = m + k \times SD$
2.576	0.50	99.5
2.326	1	99
2.06	2	98
1.96	2.5	97.5
1.88	3	97
1.65	5	95
1.28	10	90
1.04	15	85
1.00	16.5	83.5
0.84	20	80
0.67	25	75
0	50	50

If *p* is above the average, add the product to the mean:

$$p = m + k \times SD$$

Examples:

To determine 95th percentile, use *k* 1.65.

To determine 20th percentile, use *k* 0.84. For more information see the reference section.

FURTHER READING

Gupta, D. (2020). New directions in the field of anthropometry, sizing and clothing fit. In N. Zakaria & D. Gupta (Eds.), Anthropometry, apparel siz-ing and design (2nd ed., pp. 3–27). Elsevier

Hernandez-Arellano, J. L., Aguilar-Duque, J. I., & Gómez-Bull, K. G. (2018). Methodology to Determine Product Dimensions Based on User Anthropometric Data. In J. L. García-Alcaraz, G. Alor-Hernández, A. A. Maldonado-Macías, & C. Sánchez-Ramírez (Eds.), New Perspectives on Applied Industrial Tools and Techniques (First, pp. 373–385). Cham, Switzerland: Springer International Publishing. http://doi.org/10.1007/978-3-319-56871-3

Merrill, Z., Perera, S., & Cham, R. (2019). Predictive regression modeling of body segment parameters using individual-based anthropometric measurements. *Journal of Biomechanics*, *96*, 109349.

Ranger, F., Vezeau, S., & Lortie, M. (2019). Tools and methods used by industrial designers for product dimensioning. *International Journal of Industrial Ergonomics*, *74*, 102844.

Roebuck Jr, J. A. (2015, September). Fixing flaws in engineering anthropometry, ears and all. In *Proceedings of the Human Factors and Ergonomics Society Annual Meeting* (Vol. 59, No. 1, pp. 1414–1418). Sage, CA; Los Angeles, CA: SAGE Publications.

Anthropometry for Design

6

E. Nowak

6.1 INTRODUCTION

The aim of this entry is to present anthropometry as a set of measuring techniques and methods, and to prove its usefulness for the needs of design. Anthropometry originates from anthropology and is directed toward obtaining measurements of man. Anthropology is the science of man. It deals with the changability of the physical characteristics of man in time and space, and particularly with race differentiation, individuals' differentiation, ontogenesis, and phylogenesis. In the English and American approaches, anthropology embraces the complete knowledge of man and can be divided into physical (biological) anthropology and cultural anthropology. In addition, there are other types of anthropology, such as social anthropology and criminal anthropology.

Physical anthropometry is particularly useful for practical purposes. According to the accepted divisions most often used, physical anthropology can be separated into the following three basic parts:

1. *Population anthropology* (known earlier as race anthropology), which is the study of the intraspecies differentiation of man—including living conditions, history, and the present state
2. *Ontogenetic anthropology*, which is the study of the ontogenesis of man
3. *Phylogenetic anthropology*, which deals with the phylogenesis of man—that is, the origin of our species

DOI: 10.1201/9781003459767-6

Anthropology is related in its research to biology and the humanities (archeology, prehistory, psychology, and pedagogy), as well as the technical sciences. Ergonomic anthropology deals with man as the basic unit of the man—technique system and, as such, has evolved and developed in tandem with technology and engineering. Together with other disciplines (including physiology and psychology), it aims at obtaining the best conditions for this system to function.

Ergonomic anthropology makes use of the scientific output of phylogenetic, ontogenetic, and population anthropology, and as a result of the problems it concentrates on, it benefits from various sections of medicine and psychology.

An anthropologist dealing with ergonomics makes use of the classical anthropological science and develops this science for the needs of technology and engineering. Thus, he or she utilizes classical anthropometry, that is, the basic research methods applied to anthropology.

6.2 AIMS AND TASKS OF ANTHROPOMETRY

6.2.1 Measuring Methods

The basic anthropometric measurements of humans include:

- Linear measurements
- Angular measurements
- Circumferences
- Force measurements

Linear measurements include: breadth, height, and length measurements. These are measured between recognized anthropometric points. Angular measurements are carried out between planes and lines that cross the human body. Body movements in the sagittal plane are termed flexions and extensions; back and head movements in the sagittal planes are termed bending to the right and to the left; and extremities' movements are termed adductions and abductions. According to these, extremities' movements in the transverse plane are termed pronations and supinations, and back movements are termed left turns and right turns.

Circumferential measurements of the body are mainly carried out for the purposes of clothing design and for physical assessment. The basic measurements include: head, neck, chest, hips, arms, thigh, and shin circumferences. Force measurement is done in order to define the physical predispositions of man. In general, force is defined in relation to that exerted by the hand and foot. Moments of forces are used as data applied to the design of hand and foot control systems.

The basic aim of classical anthropometry is to provide objective and precise data on the somatic structure of man. In population anthropology, for example, anthropometry is used as a set of methods applied to defining biological differences that occur between human populations. Anthropometry in ontogenetic anthropology serves to assess the ontogenetic development of man, and provides data for defining the development process, the process of puberty, and the aging process.

Following the development of the body segments in particular ontogenetic periods, anthropometry describes changes in the proportions of the human body. During the physical development of humans, head measurements increase twofold, trunk measurements increase threefold, and limb measurements increase between four- and fivefold. The aim of anthropometry is not only to define differences in human body structure in relation to age, but also in relation to sex and the type of somatic structure. Significant differences can be found in somatic characteristics in men and women.

Following the development of successive generations, anthropometry assists in defining and foreseeing developmental trends of populations. These concern such phenomena as development acceleration and secular trend. These phenomena result in significant differences between generations in somatic, morphological, and functional characteristics. Anthropometry describes these differences and provides data on somatic changes that occur in given populations. Pediatrics and ergonomics are able to make use of these data: pediatrics applies population data as biological standards to the evaluation of individuals; on the basis of somatic characteristics of a given population, ergonomics creates products adjusted to the body structure. This development in ergonomics has resulted in the production of methods and measuring techniques applied to anthropometry, and new techniques aimed especially at the needs of ergonomics, termed ergonomic anthropometry, have gradually evolved.

The main aim of ergonomic anthropometry is to provide data describing physical predispositions of humans in order to aid the design of work and living environments. This objective has resulted in the modification of existing methods and the development of new ones. For example, in ergonomics many anthropometric measurements are performed on the basis of classical anthropometric points and new fixed references. In classical anthropometry, the main reference basis for height measurements is the horizontal place of footrest—**Basis** (B). An additional vertical plane basis—**Basis dorsalis** (Bd)—has been introduced to meet the needs of ergonomics. This basis is mainly applied in the determination of body dimensions in the sagittal plane. These dimensions include depth measurements and reaches. To measure the body in the sitting position, two additional reference planes have been introduced. These are the horizontal seat plane—*Basis sedilis* (Bs) and the vertical plane—*Basis sedilis dorsalis* (Bsd).

Ergonomic anthropometry has frequently developed reference systems to solve specific problems in respect of the requirements of constructors and designers of technical machines and appliances. The application of this method was used to determine the dimensions of the spatial zone of upper limb reaches (Bullock 1974; Nowak 1978).

Initially, during the development of ergonomic anthropometry, adults were the main subject since at that time ergonomics dealt with the work environment of man. As ergonomics developed, its interests were extended to the life environment of man and included home ergonomics and leisure ergonomics. As the role of ergonomics increased, it was able to benefit from the use of data that contemporary anthropometry could supply on the various stages of human development. Assuming ontogenetic periods as the criteria for division, we can distinguish anthropometry for children and young people, adults and the elderly. Anthropometry applies adequate measuring techniques for investigating each of these groups. For example, the length measurements of children up to one year old are undertaken in the prone position by means of a special type of liberometer. The same measurements of the adult population are performed in the sitting position by means of the vertical anthropometer. The majority of methods and measuring techniques applied in anthropometry can be used for measuring disabled people. However, some of these are modified or simplified in view of the difficulties in obtaining measurements. For example, special measuring chairs are constructed to study people with lower extremity dysfunction as they can be studied only in the sitting position.

It is not easy to study the disabled using the methods applied by classical anthropometry. It requires a great deal of experience from those performing the experiments since the measurements need to be taken very quickly. The measuring methods that make it possible to carry out investigations at a distance are the most convenient, both for the subject and the investigator; these are termed nontactile methods.

This type of method was applied by the Swedish researcher Thoren (1994) in carrying out measurements on a group of the disabled. A set of mirrors and cameras properly arranged and interrelated to CAD/CAM software made it possible to obtain anthropometric data in a spatial system very quickly. Photogrammetric methods are most often used in the measurements of disabled people, as these make it possible to assess deformities and changes in the body structure, and dislocations of bones segments. It is also possible to define the shape and dimensions of the body regardless of the body position and its changes in time (Das and Kozey 1994). It should be mentioned that these kinds of methods are relatively expensive and not all research units can afford to use them.

Summarizing the above, it can be stated that the sets of measuring techniques and methods of classical and ergonomic anthropometry can be applied to the measurements of the disabled. Some of these methods, however, require verification from the point of view of the arduousness of investigations.

6.3 STATISTICAL METHODS

Individuals with various body dimensions (tall and short) and body proportions (long or short extremities, or long or short trunks) exist in every population. In order to characterize a given population—that is, evaluate it in terms of numbers—anthropometry utilizes the basic statistical characteristics. Usually anthropometric features have the normal distribution and are arranged according to the Gaussian distribution. Figure 6.1 illustrates such distribution.

It presents the body height of Warsaw girls aged 18years (Nowak 1993). The values of this feature were presented on the frontal axis, and the frequency of occurring (possibility) on the horizontal axis. Two basic statistical parameters are used to determine the distribution of features in a population. One of these parameters is the mean (*m*). It indicates where the distribution is

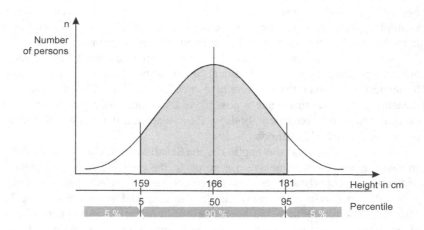

FIGURE 6.1 Distribution of body height (stature) in Polish girls (18 years of age). About 90% of all girls are between 159 and 181 cm tall; about 10% are either shorter or taller.

located on the horizontal axis. The other is a quantity known as the standard deviation (*S*), which is the index of the degree of variability in the population under study—the "width" of the distribution or the extent to which individual values are scattered about or deviate from the mean. Mean values and standard deviations are used to determine the statistical characteristics called percentile. They are useful both in developing biological standards and preparing standards for the needs of ergonomics. Assuming that the features investigated in a random test of the population have a normal distribution, as seen in Figure 6.1, it can be termed the percentile (*Cp*): the value of the feature that does not exceed *p*% of individuals. The values of particular percentiles (*Cp*) are calculated according to the following formula:

$$Cp = m + S \times z$$

where *Cp* is the characteristics value on the level of the *p* percentile; *m* is the mean; *S* is the standard deviation; *z* is the constant for the percentile concerned (see statistical tables).

In order to obtain better percentile interpretation we need to return to Figure 6.1. The height measurement of the investigated population of girls aged 18 is distributed in a symmetrical way (Nowak 2000). Its highest point is the average stature, otherwise known as the mean. Since the curve is symmetrical, it follows that 50% of the population of girls are shorter than average and 50% are taller. In this distribution, the mean is equal to the 50th percentile. Other percentile values are also marked on the horizontal axis. The 5th and 95th percentile are used for designing purposes. The 5th percentile located closer to the frontal axis means that 5% of the girls are shorter. Similarly, an equal distance from the mean toward the right of the chart is a point known as the 95th percentile. Hence, we can say that only 5% of the girls are taller. Ninety percent of the population are between the 5th and 95th percentile in stature. Thus, using the values of the 5th and 95th percentile and applying the rules of ergonomics, products for 95% of the population can be designed.

6.4 SOMATIC CHARACTERISTICS OF THE EUROPEAN POPULATION: DATA FOR DESIGNING

Globalization and European integration processes, which have intensified in the last few years, and the relating free flow of goods also impose new tasks on ergonomic anthropometry. Export sales require manufacturers to know the body measurements of the population of the country where the products are to be sold. In order to secure a ready market for the manufactured goods, every producer has to adjust them to the somatic characteristics of their potential user as early as at the stage of designing. Inadequate anthropometric data used in the design negatively influence working conditions and decrease the efficiency of use and functional values of a product. In general, each country develops sets of data to define the measurements of the native population for the particular number of features. Not all countries, however, have up-to-date collections of data prepared for the needs of design. Table 6.1 presents anthropometric data of the Polish population developed as the prognosis for the year 2010 (Nowak 2000). Table 6.2 includes the comparison—undertaken on the basis of data available in the literature—of the stature measurements of adult men and women living in the following European countries: France (Rebiffe *et al.* 1981), Great Britain (Pheasant 1996), Germany (Flügel *et al.* 1986), Holland (Peebles and Norris 1998), Italy (Masali *et al.* 1992), Norway (Bolstad *et al.* 2001), and Poland (Nowak 2000).

TABLE 6.1 Anthropometric Measured Data (in mm) of Polish Men and Women Aged between 19 and 65 (Nowak 2000)

	MEN 19–65 YEARS			WOMEN 19–65 YEARS		
	PERCENTILES					
DIMENSION	5	50	95	5	50	95
Stature	1660	1778	1890	1536	1634	1740
Eye height	1524	1650	1771	1438	1503	1615
Acromion height	1403	1459	1550	1283	1337	1431
Suprasternal height	1354	1449	1570	1238	1324	1415
Elbow height	1027	1098	1207	1012	1017	1124
Pubic height	874	926	1011	790	850	910
Head and neck height	312	324	330	298	306	325
Trunk height	474	528	549	448	478	505
Thigh length	420	436	452	344	395	432
Knee height	454	490	550	450	460	475
Upper extremities length	687	777	852	632	707	783
Arm length	291	333	374	273	303	333
Forearm length	226	259	276	203	232	262
Hand length	165	180	196	152	167	182
Arm overhead reach[a]	2053	2127	2304	1849	2004	2121
Arm reach down[a]	675	768	846	653	722	785
Arms span[a]	1456	1577	1705	1327	1439	1576

[a] Measurement taken with the hand clenched

TABLE 6.2 Stature for European Male and Female Subjects (in mm)

COUNTRY	MALE			FEMALE		
	5TH	50TH	95TH	5TH	50TH	95TH
France	1607	1719	1830	1507	1604	1705
Germany	1629	1733	1841	1510	1619	1725
Great Britain	1641	1755	1869	1514	1620	1726
Netherlands	1676	1793	1910	1564	1660	1757
Italy	1604	1728	1847	1500	1610	1710
Norway	1688	1796	1904	1561	1661	1761
Poland	1660	1778	1890	1536	1634	1740

The source data implicate that the numbers included in Table 6.2 were gathered in different years. The oldest data concern the French and German populations and were obtained in the 1980s. The latest ones describe the Norwegians and were published by Bolstad and others in the *Applied Ergonomics* journal in 2001. Data on the Polish population are the target standard up to the year 2010, and they were published by the Institute of Industrial Design in the monography entitled: *Anthropometric Atlas of the Polish Population. Data for Design* (Nowak 2000). The populations under investigation representing particular countries do not embrace subjects of the same age. The age range for the Norwegian population is to 20–39 years, for the Italian one 18–83 years, and for the British and Polish ones 19–65 years. The adult Dutch population included subjects aged over 20. The year of the investigation and the age of the population under investigation are of great significance for the results obtained. Low stature of the Italians can be caused not only by genetic and environmental factors, but also by the fact that the group of subjects included the elderly over 65 years of age. Measurements—and especially those of the stature characteristics—obtained in earlier years could have changed due to the occurrence of a secular trend, and today they would be different. In general, it is assumed that the mean stature value changes every ten years by about 1 cm. In Poland, as in other countries, systematic stature growth is observed, although this trend has become less intensive. Measurements taken in the years 1966–1972 showed that the mean stature was 1695 mm for Polish men and 1577 mm for women. By the year 2000, these measurements increased to 1748 mm for men and 1615 mm for women. Over 30 years, the stature measurement increased by 53mm in men and 38 mm in women. In the years 1981–1995, the

UK noted a stature growth of 17 mm in men and 12mm in women (Pheasant 1996). Bolstad *et al.* (2001) stated that in the years 1980–1994 the mean body height of Norwegian men increased by only about 3 mm and that of Swedish men by 2 mm. These differences result from the type of nourishment and hygienic and sanitary conditions of the population. It is assumed that developed countries witness the gradual extinction of a secular trend, since the balance between living conditions and the genetic potential of the population has been gained.

Despite the changes that have been occurring and will take place during the physical development of particular populations, anthropometry directed to the needs of ergonomics should create, as far as possible, up-to-date sets of anthropometric data to inform manufacturers of the body structure of their products' consumers.

Table 6.3 presents the minimum and maximum values of 25 somatic characteristics of the seven European populations mentioned earlier.

In general, the minimum measurements consist of the values of the 5th percentile of the Italian or French population, whereas the maximum ones concern the Dutch and Norwegians. This type of synthetic presentation of anthropometric data gives a designer the body dimensions of Europeans in spite of the differences occurring among the populations. Some European and world standards include common measurements for men and women despite the significant differences in the body structure of female and male individuals. This does not seem appropriate from the point of view of anthropology, nevertheless it significantly simplifies the process of data analysis and therefore is more practical for a designer. In general, women are from 8 to 15% smaller than men. Differences in stature between a short woman and a tall man be as much as 150–350 mm. In spite of such significant differences, some of the authors present anthropometric data on men and women calculated jointly for both sexes. Joint measurements of men and women are included in the Prestandard DIN-EN-ISO 15537:2002: Principles for selecting and using test persons for testing anthropometric aspects of industrial products and design (ISO/DIS 15537:2002). The Prestandard presents data of European and world populations. Global data include both smaller and larger types. Data concerning stature measurements are presented in Table 6.4.

In order to adjust products to the anatomic structure of man at every stage of his or her ontogenetic development, anthropometry supplies designers with data concerning the body structure of not only adults, but also that of children, juveniles, the elderly, and the disabled. The above mentioned data can be found in the relevant entries of this encyclopedia.

TABLE 6.3   Anthropometric Data of the European Male and Female Subjects (in mm)

MEASUREMENT	MALE		FEMALE	
	MIN (C5)	MAX (C95)	MIN (C5)	MAX (C95)
Stature	1604	1910	1507	1757
Eye height	1450	1765	1400	1765
Shoulder height	1300	1591	1210	1465
Elbow height	973	1215	905	1145
Shoulder breadth	420	588	355	525
Chest breadth	250	330	233	306
Chest depth	180	330	148	350
Hip breadth	296	443	296	478
Arm reach down (to grip)	745	845	655	845
Arm reach forward (to grip)	631	898	616	836
Arm overhead reach (to grip)	1835	2304	1649	2121
Sitting height	830	995	792	930
Eye height, sitting	711	875	644	805
Shoulder height, sitting	545	671	515	631
Elbow height, sitting	175	301	165	292
Popliteal height, sitting	383	521	350	481
Popliteal depth, sitting	429	588	422	555
Thigh thickness, sitting	115	211	106	195
Head length	175	213	165	197
Head breadth	136	170	130	177
Hand length	165	210	152	195
Hand breadth	75	100	68	90
Foot length	235	295	215	264
Foot breadth	85	112	80	107
Body mass (in kg)	58	101	45	88

TABLE 6.4   Values of the Stature for the European and World Populations Following

	5TH	50TH	95TH
European population World population	1530	1719	1881
Smaller type	1390	1520	1650
Bigger type	1650	1780	1910

6.5 APPLICATION OF ANTHROPOMETRY

Examples of the applications of anthropometric investigations are given below; they are designs for work spaces, handles and holders, and working clothes.

Work paces determined mainly by the reach zone of the upper extremities defined in relation to three planes: sagittal, transverse, and frontal. The first attempts to define the upper extremities reach zone concerned one- or two-dimensional configurations determined in frontal or transverse planes. Barnes (McCormick 1964) defined the upper extremities reach zone by determining the so-called maximum and normal zones. These zones were determined by the radii, which constituted the length of the arm or forearm. Further investigations aimed at determining the reach zone concerned three-dimensional configuration. Reach zones were defined on the basis of experimental research. Research work conducted in the USA (Damon *et al.* 1966) and Australia (Bullock 1974) should be mentioned here. The results of both experiments provided data for military purposes. The results of investigations conducted in Poland (Nowak 1978), thanks to the fact that the measuring system was "suspended" on acromial points, can be used while designing all kinds of work stands intended for work in both the standing and sitting positions. Work space of the upper extremities was defined based on investigations of 226 men and 204 women aged from 18 to 65 years (Nowak 1978).

In these investigations, a spatial measuring system consisting of three interperpendicular axes is assumed to determine the arm reach area. The measuring system of the arm reach area was determined by the intersection of the following planes: the frontal plane, tangential to the vertical plane of the seat back, the sagittal-median plane, the transverse plane crossing the acromial points.

The intersection of these planes marks the origin of the polar co-ordinate axes of the measuring system (point C). The C point is fundamental to define the reach areas for any working plane. The whole reach space of arm reach was divided into ten horizontal measuring planes. That one which crosses the acromial points was accepted as the basis and marked O. From this plane upwards four others follow every 120 mm and five follow downwards. The reach ranges of the left and right arms were recorded on each plane in a polar system with its center being the C point.

The investigations resulted in ten arm reaches for the left and right extremities. These reaches determined in a polar system define the spatial area of arms. This space is a basis for designing spatial structures of machines, installations, and workplaces.

Figure 6.2 shows an example of data being used on the reach area of arms in the design process of a real workplace situation.

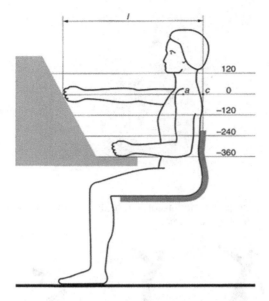

FIGURE 6.2   An example of using arm reach area data in the design process.

For the correct design of all kinds of handles and holders, it is necessary to determine the functional capabilities of the hand and its anthropometric dimensions. The human hand is a specific and highly precise work tool. The grip capabilities of the hand are supported by the opposition of the thumb. This enables the hand to perform a wide range of manipulation tasks. Almost all types of activities require different hand arrangements, which means using different types of grips. The performance of a handle—that is, its shape and form—should be adjusted to the most convenient grip to perform a specific activity. There are many grip classifications. One of them assumes as a criterion of division the form of hand arrangement or its parts in relation to the object and direction of effective force used (Nowak 2004) (Figure 6.3). Following Nowak (2004), this is a hand clamp grip making use of the thumb opposition. According to other authors, this grip is defined as a cylindrical or coiled grip. This grip is applied while using hand tools including saws, drilling machines, or hammers, as well as when opening a door with the use of a door handle. When we know the dimensions of the space occupied by the hand grasped on a cylindrical handle and the handle diameter, we can easily determine the distance of a door handle in relation both to the door plane and the door frame.

Anthropometric measurements taken for the purpose of clothes design embrace a set of characteristics different from measurements made for the purpose of designing work spaces, machines, and tools. First of all, measurements

FIGURE 6.3 Hand grip named according to different classifications: power grip, hammer grip, or clench grip palmar.

FIGURE 6.4 Functional anthropometric measurements for the needs of clothes design.

of the following circumferences are needed: neck, chest, waist, hip, wrist, and thigh. For the design of working clothes, angular measurements of the extremities' movement ranges are essential, as well as measurements defined as arcs: for example, the arcs of the front length of the trunk, the length of the back, shoulders, upper and lower extremities are all very important. The dimensions of the arcs are determined in a motionless standing position and while changing body position—for example, moving to squatting, kneeling, and standing positions, for the latter, both bending forwards and bending backwards. Additionally, the construction of garments for wheelchair users is based on the data obtained in the sitting position with consideration given to the trunk movement forward and movements of the lower extremities. Figure 6.4 presents 11 functional anthropometric measurements taken for the needs of clothing design (Batogowska 1976).

ACKNOWLEDGMENTS

I would like to express my gratitude to my colleagues and friends from the Institute of Industrial Design in Warsaw for their assistance in preparing the materials for this work. Without their help I would not be able to prepare the above chapter so quickly and efficiently. I am also thankful to the IWP research teams, in charge of which I had the honor to be, for their commitment and assistance in the work the results of which I can present to you in this book.

REFERENCES

BATOGOWSKA, A., 1976, Dynamic anthropometric measurements for the needs of design. *Prace I Materialy IWP* (Warsaw: Institute of Industrial Design) (in Polish).

BOLSTAD, G., BENUM, B. and ROKNE, A., 2001, Anthropometry of Norwegian light industry and offices workers. *Applied Ergonomics*, 32(3), 239–246.

BULLOCK, M., 1974, The determination of functional arm reach boundaries for operation of manual controls. *Ergonomics*, 3, 375.

DAMON, A., STOUDT, H.W. and McFARLAND, R.A., 1966, *The Human Body in Equipment Design* (Cambridge, MA: Harvard University Press).

DAS, B. and KOZEY, I., 1994, Structural anthropometry for wheelchair mobile adults. In *12th Triennial Congress of International Ergonomics Association*, vol. 3, Toronto, p. 63.

LÜGEL, B., GREIL, H. and SOMMER, K., 1986, *Anthropologischer Atlas* (Berlin: Verlag Tribüne).

MASALI, M.I., CONIGLIO, E., FUBINI, C., MASIERO, G., PIERLORENZI, A., MILLEVOLTE, A. and RICCIO, G., 1992, Anthropometric characteristics of Italian population from an ergonomic aimed research (unpublished). In *Proceedings of the International Congress of European Anthropological Association*, Madrid.

McCORMICK, E.J., 1964, *Human Factors Engineering* (New York: McGraw-Hill).

NOWAK, E., 1978, Determination of the spatial reach area of the arms for workspace design purposes. *Ergonomics*, 7, 493.

NOWAK, E., 1993, Anthropometric data for designing a pupil's workstand. *Prace I Materialy* IWP (Warsaw: Institute of Industrial Design) (in Polish)

NOWAK, E., 2000, *Anthropometric Atlas of the Polish Population — Data for Design* (Warsaw: IWP).

NOWAK, E., 2004, Functional assessment of a child's hand for the needs of ergonomics and rehabilitation. *Ergonomia IJE&HF*, 26(3), 227–252.

PEEBLES, L. and NORRIS, B., 1998, *ADULTDATA: The Handbook of Adult Anthropometric and Strength Measurements — Data for Design Safety* (Nottingham: Institute for Occupational Ergonomics, University of Nottingham).

PHEASANT, S.T., 1996, *Bodyspace: Anthropometry, Ergonomics and Design* (London: Taylor & Francis).

REBIFFE, R., GUILLEN, J. and PASQUET, P., 1981, *Enquete Anthropometrique sur les Conducteurs Francais* (Unpublished) (Paris: Laboratoire de Physiologie et de Biomecanique de l'association Peugeot-Renault).

THOREN, M., 1994, Clothing made to fit the disabled users. In *12th Triennial Congress of International Ergonomics Association,* vol. 3, Toronto, p. 187.

FURTHER READING

Joshi, M., & Deshpande, V. (2019). A systematic review of comparative studies on ergonomic assessment techniques. *International Journal of Industrial Ergonomics*, 74, 102865.

Kaya, Ö., & Özok, A. F. (2017). The importance of anthropometry in design. *Journal of Engineering Sciences and Design*, 5, 309–316.

Kharb, S. S., Belokar, R., & Kant, S. (2017). Study of the role of Anthropometry in Designing. *Journal of Mechanical and Mechanics Engineering*, 3(1), 2.

Molenbroek, J. (2019). DINED-Anthropometry in design. *TU Delft, [Online]*. Available: www. dined. nl.

Zare, M., Croq, M., Hossein-Arabi, F., Brunet, R., & Roquelaure, Y. (2016). Does ergonomics improve product quality and reduce costs? A review article. *Human Factors and Ergonomics in Manufacturing & Service Industries*, 26(2), 205–223.

Ergonomic Workstation Design

7

B. Das

7.1 INTRODUCTION

Ergonomic workstation design based on engineering anthropometry and occupational biomechanics can play a major role in the reduction of many risk factors of occupational injury (Grandjean 1982). Anthropometry and biomechanics are closely related, because occupational biomechanics provide the bases for the use of engineering anthropometry to the problems of workstation design (Pheasant 1986).

Anthropometry is the technology of measuring various human physical straits, primarily such factors as size, mobility and strength. Engineering anthropometry is the effort in applying such data to workstations, equipment, tools and clothing design to enhance the efficiency, safety and comfort of the worker. In the context of the workstation design, engineering anthropometry is employed to develop design parameters or dimensions for such a design.

Biomechanics deals with various aspects of physical movements of the human body and body members. The human body is considered as a linked body segments subjected to the internal forces generated within the musculoskeletal system and external forces imposed by the work situation (Karwowski 1992). Biomechanics uses law of physics and engineering concepts to describe various body parts and the forces acting on these parts during work activities.

DOI: 10.1201/9781003459767-7

The biomechanics or anatomy applied to the work situations is often considered the scientific basis of ergonomics (Tichauer 1975). Occupational biomechanics is defined as the physical interaction of workers with their tools, machines and materials so as to enhance the workers' performance while minimizing the risk of musculoskeletal disorders (Chaffin and Andersson 1984).

An ergonomically designed workstation attempts to obtain an adequate balance between worker capabilities and work requirements. The objective is to optimize worker productivity and the total system and at the same time enhance worker physical and mental well-being, job satisfaction and safety. Often workstation in industry is designed in an arbitrary manner with little attention to anthropometric measurements and biomechanical considerations of the anticipated user. The situation is aggravated by the lack of usable design parameters or dimensions (Das and Grady 1983a). It is essential to understand the conditions under which biomechanics can be used for the assessment of a workstation (Marras 1997). When the magnitude of loads imposed upon the body is believed to be exceeding the tolerance of a structure, a biomechanical analysis is considered most helpful.

7.2 WORKSTATION DESIGN BASED ON ENGINEERING ANTHROPOMETRY

In a workstation design an attempt is made to achieve an optimum compromise between the variable anthropometry of the targeted operator population, and the physical size and the layout of the workstation components. An ergonomic analysis for a workstation design is concerned with spatial accommodation, posture, reaching abilities, clearance and interference of the body segments, field of vision, available strength of the operator and biomechanical stress. The appropriate anthropometric data regarding body size, strength, segment masses and inertial properties from the established databases are typically used in the analysis.

Das and Grady (1983a, b) determined workstation design dimensions through the use of the existing anthropometric data, so that these dimensions can be readily employed by a designer. Workspace design dimensions were determined for industrial tasks in sitting, standing and sit–stand positions. Worker populations consisted of a combination of male and female workers and the individual male and female workers for the 5th, 50th and 95th percentiles based on existing anthropometric data. The normal and maximum reach dimensions were based on the most commonly used industrial operations, which require a grasping movement or thumb and forefinger manipulations. However, appropriate allowances were provided to adjust reach dimensions for other types of industrial operations. The normal and maximum horizontal and vertical clearances and reference points for the horizontal and vertical clearances were established to facilitate the design. The concepts developed by Farley (1955) and Squires (1956) were used to describe the workspace envelope for the individual worker. The dimensions of smaller (5th percentile) and larger (95th percentile) workers were used to determine the limits of reach and clearance requirements respectively.

7.2.1 Anthropometric Data Adjustment

The existing anthropometric data were derived on the basis of the measurements from nude subjects. Therefore the data were adjusted for clothing and shoe allowances. Since the data for the standing, sitting and eye heights were based on erect positions at work or rest, the data were adjusted to account for the "slump" posture involved in the "normal" standing and sitting positions. Also, necessary adjustments were made for the reach dimensions used for performing various industrial operations (Hertzberg 1972). The corrected or adjusted anthropometric measurements to account for clothing, shoe and slump posture are presented in Table 7.1 (Das and Grady 1983a).

TABLE 7.1 Corrected Anthropometric Measurements to Account for Clothing, Shoe and Slump Posture (Das and Grady 1983a)

		PERCENTILES (CM)		
SEX	BODY FEATURE	5TH	50TH	95TH
Male	Total height (slump)	166.2	176.1	186.3
	Body height (sitting, slump)	82.1	87.6	92.7
	Eye height (slump)	160.0	164.9	174.8
	Eye height (sitting, slump)	70.9	76.2	81.3
	Shoulder height	138.0	147.7	156.8
	Shoulder height (sitting)	54.7	59.8	64.4
	Body depth	26.7	30.2	34.0
	Elbow-to-elbow	40.0	45.1	51.7
	Thigh clearance	14.2	16.2	18.5
	Forearm length	37.1	40.4	43.7
	Arm length	68.3	75.2	82.0
	Elbow height	105.6	113.0	120.4
	Elbow height (sitting)	18.8	23.1	27.4

(Continued)

TABLE 7.1 (Continued)

SEX	BODY FEATURE	PERCENTILES (CM)		
		5TH	50TH	95TH
	Popliteal height (sitting)	42.4	45.7	48.7
Female	Total height (slump)	153.8	163.2	173.8
	Body height (sitting, slump)	78.5	82.8	87.1
	Eye height (slump)	143.6	153.5	163.4
	Eye height (sitting, slump)	68.6	72.4	76.5
	Shoulder height	127.9	137.5	146.6
	Shoulder height (sitting)	52.2	57.2	61.8
	Body depth	21.8	24.6	27.6
	Elbow-to-elbow	35.7	38.2	43.8
	Thigh clearance	12.4	14.4	16.5
	Forearm length	32.5	36.6	40.7
	Arm length	60.2	66.0	72.4
	Elbow height	99.0	105.1	111.2
	Elbow height (sitting)	18.8	23.1	27.4
	Popliteal height (sitting)	37.3	40.6	43.6

7.2.2 Workstation Design Parameters

To design a workstation, it is necessary to obtain relevant information or data on task performance, equipment, working posture and environment. In the case of a new workstation design, it is advantageous to obtain such information from a similar task/equipment situation. Before redesigning an existing workstation in industry, often it is desirable to conduct a worker survey through appropriate questionnaires to determine the effect of the equipment or system design on employee comfort, health and ease of use (of the equipment). However, frequently it is necessary to design a new industrial workstation. Even then, it may still be desirable to obtain feedback from the operators, who are engaged in performing a similar type of industrial task. The feedback may generate heightened awareness of workstation design problems and issues.

In the beginning, decisions are formalized regarding the task sequence, available space, equipment and tools. Work method needs to be established before embarking on the design of the workstation. Determination of the workstation dimensions usually proceeds according to the steps outlined in Table 7.2 (Das and Sengupta 1996).

TABLE 7.2   A Systematic Approach for the Determination of Workstation Design Parameters (Das and Sengupta 1996)

1. Obtain relevant information on the task performance, equipment, working posture and environment through direct observation, video recording, and/ or input from experienced personnel
2. Identify the appropriate user population and obtain the relevant anthropometric measurements or use the available statistical data from anthropometric surveys
3. Determine the range of work height based on the type of the work to be performed. Provide an adjustable chair and a foot rest for a seated operator and an adjustable work surface or platform for a standing operator
4. Layout the frequently used hand tools, controls, and bins within the normal reach space. Failing that, they may be placed within the maximum reach space. Locate control or handle in the most advantageous position, if strength is required to operate it
5. Provide adequate elbow room and clearance at waist level for free movement
6. Locate the displays within the normal line of sight
7. Consider the material and information flow requirements from other functional units or employees
8. Make a scaled layout drawing of the proposed workstation to check the placement of individual components
9. Develop a mock-up of the design and conduct trials with live subjects to ascertain operator–workstation fit. Obtain feedback from the interest groups
10. Construct a prototype workstation based on the final design

The workstation design procedure commences with the collection of relevant data through direct observation, video taping and input from experienced operators and supervisors (step 1, Table 7.2). It is necessary to identify the appropriate user population based on such factors as ethnic origin, gender and age (step 2). The necessary anthropometric dimensions of the population are obtained or approximated from the results of the available anthropometric surveys that reasonably represent the user group. As these dimensions are taken from nude subjects in erect posture, they need to be corrected appropriately for the effect of clothing, shoe and normal slump posture during work (Das and Grady 1983a).

In developing an industrial workstation, a designer should take into account the workstation height (step 3). The work table height must be compatible with worker height, whether standing or sitting. The best working height is ~2.5 cm below the elbow. However, the working height can vary several centimeters up or down without any significant effect on performance. The nature of the work to be performed must be taken into consideration in determining work height. For seated operators, provide adjustable chair and foot rests; for standing operators, provide an adjustable work surface or platform.

The hand tools, controls and bins that are frequently used need to be located within the normal reach spaces (step 4). The items used occasionally may be placed beyond the normal reach, but they should be placed within the maximum reach space. For locating a control that requires strength, give consideration to the human strength profile in the workspace. Extreme reach space, involving twisting of trunk, ought to be avoided at all times. Adequate lateral clearance must be provided for the large (95th percentile) operator for ease of entry and exit at the workstation and also to provide ample elbow room for ease of work (step 5).

The placement of the displays should not impose frequent head and/or eye movement. The optimum display height for the normal (slump) eye height is 15° downward gaze (step 6). Appropriate personnel from other functional units or departments should be consulted regarding material and information flow requirements (step 7). It is beneficial to consider the physical size of the individual components and make a scaled layout drawing of the proposed workstation to check the placement of the individual components within the available space (step 8). The operator–workstation fit should be evaluated with a workstation mock-up and through an appropriate user population (step 9). This will ensure that the task demand and layout do not impose an undesirable working posture. It is desirable to check the interference of body members with the workstation components. If necessary, the design should be modified. Finally, it is beneficial to construct a prototype workstation based on the final design. All the ten steps shown in Table 7.2 may not be applicable in every industrial workstation design situation.

7.2.3 Workstation Dimensions

For the physical design of industrial workstations, the four essential design dimensions are: (1) work height, (2) normal and maximum reaches, (3) lateral clearance and (4) angle of vision and eye height.

7.2.3.1 Work Height

Height of the working surface should maintain a definite relationship with the operator's elbow height, depending upon the type of work. The standing work heights for the 5th, 50th and 95th percentile female operators for performing different types of work for US population are presented in Table 7.3. The table provides guidelines especially for the design of delicate, manual and forceful work. Similar data for males can be obtained from Ayoub (1973) and Das and Grady (1983a).

TABLE 7.3 Standing Work Surface Height for Female Operators in Centimeters Parameters (Das and Sengupta 1996)

	POPULATION PERCENTILE		
TYPE OF WORK	5TH	50TH	95TH
Delicate work with close visual requirement	99–104	110–115	116–121
Manual work	84–89	90–95	96–101
Forceful work aided by upper body weight	59–84	65–90	71–96

7.2.3.2 Normal and Maximum Reaches

The tip of the thumb defines the normal reach while the forearm moves in a circular motion on the table surface. During this motion, the upper arm is kept in a relaxed downward position. The "maximum" reach can be considered as the boundary on the work surface in front of an operator to which s/he can reach without flexing his/her torso. For performing repetitive tasks, the hand movement should preferably be confined within the normal working area. The controls and items of occasional use may be placed beyond the normal working area. Nevertheless, they should be placed within the maximum working area.

The concept of normal and maximum working areas (Das and Grady 1983b; Das and Behara 1995) describes the working area in front of the worker in a horizontal plane at the elbow level; the areas are expressed in the form of mathematical models. The most frequently used area of the workstation preferably should be within the normal reach of the operator. The reach requirements should not exceed the maximum reach limit, to avoid leaning forward and bad posture. The maximum working area at the elbow level is determined from the data provided in Table 7.4. The adjusted anthropometric measurements for the arm length (K), shoulder height (E), elbow height (L), which are used to calculate arm radii (R) for the 5th, 50th and 95th percentiles for females.

7.2.3.3 Lateral Clearance

A well-known approach is to design the reach requirements of the workstation corresponding to the measurements of the 5th percentile of the representative group and the clearance corresponding to the 95th percentile measurements, so as to make the workstation compatible for both small and large persons. The minimum lateral clearances at waist level are determined by adding 5 cm on both sides or 10 cm to hip breadth (standing). Considering the elbow to elbow distance and the sweep of both the elbows within the normal horizontal working area and adding 5 cm on both sides, minimum lateral clearance at elbow level is determined. The values for lateral clearances are shown in Table 7.5.

TABLE 7.4   Anthropometric Measures for Females and Maximum Reach in Centimeters Parameters (Das and Sengupta 1996)

POPULATION PERCENTILE	ARM LENGTH (K)	SHOULDER HEIGHT (E)	ELBOW HEIGHT (L)	MAXIMUM REACH (R)
5th	60	128	99	53
50th	66	138	105	58
95th	72	147	111	63

TABLE 7.5   Anthropometric Measurements for Females for Lateral Clearances in Centimeters Parameters (Das and Sengupta 1996)

POPULATION PERCENTILE	HIP BREADTH	ELBOW TO ELBOW	CLEARANCE BODY DEPTH	CLEARANCE WAIST LEVEL C1W10	CLEARANCE ELBOW LEVEL C2HG10
5th	40	36	22	50	68
50th	45	38	25	55	73
95th	52	44	28	62	81

7.2.3.4 Angle of Vision and Eye Height

Das and Grady (1983a, b) have provided the eye height for the standing female operators: 143.6 cm for 5th percentile, 153.5 cm for 50th percentile and 163.4 cm for 95th percentile. For males, similar eye height data can be obtained from the source stated above. Using trigonometry, the angle of sight can be calculated from the horizontal distance of the display from the operator's eye position.

7.3 SUPERMARKET CHECKSTAND WORKSTATION: A CASE STUDY

The steps outlined in Table 7.2 were applied to design a supermarket checkstand workstation. The working posture and work methods of the cashiers were recorded through direct observation (step 1, Table 7.2). Based on the direct observation, the major shortcomings in the design of the supermarket checkstand workstation were identified.

An operator (cashier, all female) survey was conducted in three superstores, through questionnaires to determine the effect of (1) environmental factors, (2) general fatigue induced by the task, (3) physical demand of tasks and (4) the postural discomfort of the operators during the course of a regular working day (step 2). The dimensional compatibility between the cashiers structural anthropometry and the workspace dimensions was evaluated using a scaled drawing of the existing checkstand, which was prepared from the actual measurements taken at the site.

By employing an engineering anthropometry approach, the supermarket checkstand workstation design parameters or dimensions were determined for: (1) optimum work height, (2) normal and maximum reaches, (3) lateral clearance and (4) angle of vision and eye height (steps 3–6). The appropriate data were used subsequently for the design of the checkstand workstation (Figure 7.1; Das and Sengupta 1996).

The superimposition of the normal and maximum horizontal and vertical working areas on the supermarket checkstand drawing facilitated the design or placement of the checkstand components. This procedure enabled placement of the components within the normal working area when possible and, failing that, within the maximum working area. Several alternative checkstand layout drawings were considered using a CAD package with human modeling capability. An adjustable and padded floor platform was provided for the cashiers. The platform height could be lowered by 6 cm at a time to accommodate taller cashiers. The padding was provided to reduce foot fatigue from prolonged standing.

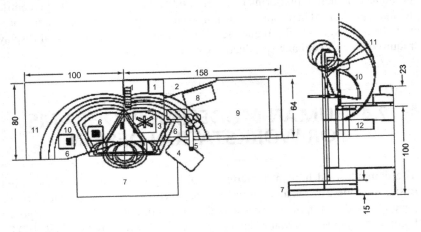

Legend:
1. Printer & code catalogue
2. Deflector
3. Laser scanner & scale
4. Keyboard
5. Price display
6. Bag hanger
7. Platforms
8. Grocery item
9. Conveyer working area
10. Normal working area
11. Maximum working area
12. Cash box

FIGURE 7.1   Normal and maximum working areas of the 5th, 50th and 95th percentiles for the female in the horizontal and vertical planes for a supermarket checkstand workstation. All dimensions in centimeters parameters (Das and Sengupta 1996).

The components that needed frequent operation were placed, as far as possible in the front of the cashier to reduce the twisting of the torso and neck. The width and depth of the laser scanner was reduced to accommodate the printer and code catalogue right in front of the cashiers. The reduced width scanner improved the cashiers' reach over both conveyor belt and the bag handling area. A deflector was provided on the conveyor belt to ensure that the products are within the maximum reach of the cashiers. The original product bin handling requirements at the left of the cashiers was eliminated and was replaced by plastic bag hangers at appropriate height and location. The superimposition of the normal and maximum horizontal and vertical working areas on the checkstand drawing facilitated the determination of the size and the placement of the checkstand components (step 8).

A time study of the simulated cashiers' task in the laboratory showed a 15% improvement in worker productivity. The main improvements in the proposed design were: (1) forward facing work posture and elimination of the requirement of twisting the torso (forward facing location of laser scanner, weigh scale, bag hangers, keyboard and printer/code catalogue), (2) increased area on the conveyor belt within the normal working area, (3) placement of the visual display (item price) within the normal line of sight of the operator and (4) adjustable height platforms to accommodate 5th, 50th and 95th percentile female operators. The proposed design would improve working posture, provide flexible work height, reduce reach requirements, improve visual display requirements and enhance productivity of the cashiers.

7.4 HUMAN MODELING PROGRAMS FOR WORKSTATION DESIGN

The computerized human modeling programs for an industrial workstation design provide a convenient computer interface for the user to interactively generate and manipulate true-to-scale, three-dimensional (3D) image of human and the workstation graphically on the video display terminal (VDT). Through the use of the programs, the designer can construct a large number of anthropometric combinations to represent the human. The programs give a complete control to the user over the development of the human model and provide a comprehensive package to evaluate human–machine interaction through easy to understand programmed commands. The user need not be a computer specialist to use such programs.

To illustrate the current state of development, Das and Sengupta (1995) had selected six representative programs:

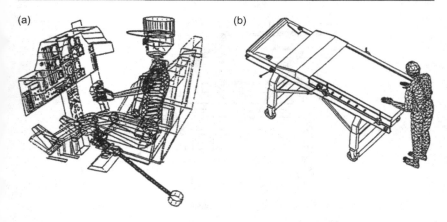

FIGURE 7.2a,b (a) JACK human model seated within a workstation (Badler 1991). (b) MANNEQUIN-generated human model in front of a meat-cutting workstation (HUMANCAD 1991).

CYBERMAN, COMBIMAN, CREN CHIEF, JACK, SAMMIE and MANNEQIN. The individual programs were compared under four broad criteria: (1) usability in terms of hardware and software; (2) anthropometry and structure of the human model; (3) model manipulation, reach and visual analysis functions and (4) other ergonomic evaluative functions (Das and Sengupta 1995). The programs differ considerably in terms of system requirement, operating characteristics, applicability and the various ergonomic evaluation functions available in the human modeling programs. For the purpose of illustration two representative models are presented (Figures 7.2a, b).

7.5 RECENT DEVELOPMENTS IN WORKSTATION

7.5.1 Design Based on Engineering Anthropometry

A computerized potentiometric measurement system (CPMS) was recently developed for anthropometric measurements of the three-dimensional maximum reach envelope (MRE) (Das *et al.* 1994). The system uses four potentiometric units, a power supply unit, a computer with analogue/ digital (A/D) converter to measure the position of a movable stylus in three

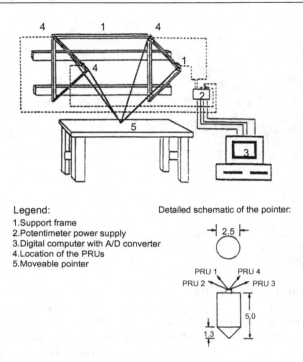

FIGURE 7.3   Schematic diagram of the computerized potentiometric system for structural and functional anthropometric measurements. All dimensions are in cm (Das *et al.* 1994).

(S, Y and Z) dimensions (Figure 7.3). This system took only 15 min for reach data collection of a participant in seated and standing positions. It recorded arm reaches while the arm was in motion, thus reflected the true dynamic nature of functional reach. The previous mechanical measurement systems recorded maximum reach envelope in terms of static arm reaches and took 3 h to collect data for the seated maximum reach envelope. The CPMS was successfully used to determine MRE for the 5th, 50th and 95th percentile males and females in seated and standing work positions (Sengupta and Das 1999).

An improved anthropometric model of 3D maximum reach envelope (MRE) was recently developed (Sengupta and Das 1998). The characteristics of the model included: application of nonlinear optimization, use of relevant structural body dimensions, use of direct measurement of dynamic reach profiles and location of the spherically shaped MRE near the shoulder joint (acronium point).

7.5.2 Workstation Design Based on Occupational Biomechanics

A biomechanical approach to workstation design deals with the effects of exertion-related forces and awkward postures in a work situation that may cause pain or injury. For the design of a workstation it is necessary to know what a person can or cannot do. A person's physical capabilities, especially those that allow an individual to exert force or sustain external loading without causing personal injury, can be determined by measuring human strengths (Mital and Das 1987). It is necessary to conduct a biomechanical evaluation of a workstation design when a worker is believed to be exposed to excessive physical stress or risk in a workstation. Through redesign of a workstation, it is possible to minimize or eliminate work-related injuries to musculoskeletal system. The low back pain is considered to be the most costly and prevalent work related musculoskeletal disorder. According the National Council on Compensation Insurance (USA), 33% of all workers' Compensation payments are due to low back pain. The total cost estimates range from US$27 to 56 billion in the USA when indirect costs are taken into consideration (Andersson *et al.* 1991).

7.5.3 Determination of Human Strength for Workstation Design

Human strengths are classified as isometric or static and dynamic. The dynamic strengths are further subdivided into: (1) isokinetic, (2) isotonic and (3) iso-intertial. In the case of isometric or static muscle exertions, the body segment involved and the object held remain stationary, while in the case of dynamic muscular exertions both the body segment and object move. Because there is no effective limb–object–muscle movement in the case of isometric strengths, such strengths do not account for the inertial forces. Consequently, isometric strengths cannot be used for the determination of an individual's capability to perform dynamic tasks, such as materials handling. Thus, for the determination of a person's physical capabilities, dynamic strengths measurement is more appropriate than isometric or static strengths measurement. For designing industrial jobs and workstations, dynamic strengths should be used even though they are difficult to measure compared to static strengths. Table 7.6 shows isometric and isokinetic strengths of males in different postures (Mital *et al.* 1986).

Industrial workers should not generally exceed one-third of their isometric strength on a sustained basis in task performance (Putz-Anderson 1994).

TABLE 7.6   Isometric and Isokinetic Lifting Strengths of Males in Different Postures (Mital *et al*. 1986)

POSTURE	STRENGTH (KG)		
	EXERTION	MEAN	STANDARD DEVIATION
Standing erect, fore arms horizontal, upper arms vertical	Isometric	40.4	6.7
Standing erect, upper arms horizontal, fore arms vertical	Isometric	52.6	8.5
Lifting floor to 81 cm height in free style	Isokinetic	110.4	30.2
Lifting from 81 cm to 152 cm in free style	Isokinetic	75.5	33.4
Pulling at 81 cm height from left to right	Isokinetic	35.1	9.5

Overloading of muscles should be avoided to minimize fatigue. Dynamic forces should be kept <30% of the maximum force that the muscle can exert; up to 50% is all right for up to 5 min. Static muscular load should be kept <15% of maximum force that the muscle can exert. General guidelines suggest that hand forces should not exceed 45 Newtons. On the other hand, it is possible to handle a force of 4kg for 10 s, 2kg for 1 min and one-third of maximum force for 4min.

For optimum design of a workstation, it is important to determine human strength profiles in the workspace. The ideal industrial workstation should be compatible with not only systems performance requirements but also with the user. The most obvious criteria are comfort and ease of use, but other equally important design criteria include work performance, safety and health (Das and Sengupta 1996). Several factors impinge upon the creation of the ideal workstation, one of which is reach capability, as discussed earlier under workstation design based on engineering anthropometry. Accurate reach capability data are essential to ensure that all hand-operated controls or tasks are located where they can be reached and operated efficiently. Another factor that impinges upon the creation of the ideal workstation is user strength capability. To ensure optimal workplace layout, it is imperative that operator's strength profile be determined. The strength profile of a person under specified conditions is essential for the design of tools (e.g. their weight, ease of use), controls (e.g. type of grip required, spatial placement) and equipment—in other words, the workstation. Furthermore, for the selection or job placement of workers requiring strength exertion in task performance, the measurement of strength profiles of such individuals can be useful.

Studies have shown that horizontal distance and vertical height of exertion significantly affect the force exertable both in static and dynamic strength tests (Chaffin and Park 1973; Davis and Stubbs 1977). However these studies have not attempted to relate anthropometric reach space envelopes to the strength data obtained. Researchers have measured strength at varying elbow angles (Hunsicker 1955), fractions of mean reach for the population (Davis and Stubbs 1977; Kumar 1991), or fixed distances (Mital and Faard 1990). Individual functional reach regions have not determined measurement locations. For optimum workstation design a link must be established between an individual's ability to reach and exert force at functional reach regions. Recently isometric push and strength profiles were determined for the able bodied population in the normal, maximum and extreme workspace reach envelopes (Das and Wang 1995). Also, isometric push, pull, push-up and pull-down strength profiles were determined for the paraplegics in the similar workspace (Das and Forde 1999; Das and Black 2000). Research is in progress at Dalhousie University, Canada to determine a comprehensive database for both static (isometric) and dynamic strength profiles in the workspace.

Insufficient physical capability while performing manual materials handling activities and tasks requiring hand tool usage can lead to overloading the muscle–tendon–bone–joint system and possible injury (Ayoub and Mital 1989). These two activities account for ~45% of all industrial overexertion injuries. It accounts for billions of dollars in worker's compensation cost (Waters and PutzAnderson 1996).

7.5.4 Posture Analysis for Workstation Design

A common manifestation of back injury is low back pain (LBP). Epidemiological studies show a positive correlation between the exposure to mechanical overload at work and the incidence rate of LBP (Chaffin and Park 1973). However, except for traumatic and acute cases, the cause of the LBP still remains unclear (Kroemer *et al.* 1996).

For the manual material handling (MMH) tasks, it is believed that the source of LBP can be traced back to the repetitive over-exertion at the lower back. This produces microtrauma at the lower spine structure over a prolonged period of time. The microtrauma ultimately results in a permanent or temporary damage to the fibrocartilegnious disks and its surround structures and can cause LBP.

An important predictor of such structural failure is the mechanical forces acting at the lumber spine which depend on the interaction of the worker anthropometry and work characteristics. In the evaluation of workstation and work method, it is important to determine the stresses at lower back due to work performed at different trunk postures. From a biomechanical perspective, the compressive force generated at lower back (L5/S1) is believed to be the most significant factor in the development of LBP.

Safe limits of work related to heavy exertion have been well established for lifting and lowering type of manual material handling (Waters and Putz-Anderson 1996) and push–pull type of exertion (Snook and Ciriello 1991). But limits for work-related lower back stresses for postural loads are not well defined. Awkward postures are of major concern for workers who are performing repetitive jobs due to the frequency and cumulative effects of exposure. Nonneutral back postures such as flexion, lateral bending, and/or twisting increase the level of muscle fatigue and intradiscal pressures in the lumbar spine (Chaffin 1973; Andersson et al. 1977). Severe trunk postures can elevate the compressive force at low back, even though a load in the hands does not exist or is relatively light in weight (Chaffin and Andersson 1991).

Several whole body posture-recording schemes have been developed (Juul-Kristensen et al. 1997) to estimate the postural stress for various types of industrial work. Their effectiveness in quantifying the postural stress has been validated by extensive field studies (Genaidy et al. 1994). Primarily the posture-recording schemes provide a means to estimate the level of postural stress based on the severity and duration of the work postures. However, most of them do not provide the safe limits of the postural stresses and thus, they are basically useful for comparison of postural stresses before and after modification of a work-site. The Ovako Working Posture Analysis System (OWAS) (Karhu et al. 1977) is a widely used postural recording scheme and has been applied to various types of industries (Pinzke 1992). A significant relationship between the back postures as defined by OWAS and prevalence in lower back pain has been established by epidemiological analysis (Heinsalmi 1986; Burdorf et al. 1991).

7.5.5 Use of Biomechanical Methods for Workstation Design

Occupation biomechanics provide a logical basis for providing data for human performance and human tissue tolerance. Biomechanical models of lower back with varied level of anatomical details and modeling capabilities have been developed to predict the compressive force in the low back due to external loads. The complex models have dealt with passive force generation and load sharing by muscles and ligaments, asymmetric postures, and inertial effects of dynamic motions. However, at the present state of development of the models, the model

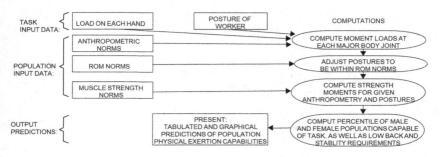

FIGURE 7.4 Biomechanical logic used to predict whole-body static exertion capabilities for given postures, hand force directions and anthropometric groups (Chaffin 1997).

outputs are still not reliable enough for judging the absolute acceptability of a work situation, rather they are more suitable for comparing back loading in different work situation (Delleman *et al.* 1992). This is because of a the inherent difficulty in substantiating the validation of the model outputs and the uncertainties of the living tissue strength values against which the model predictions would be compared. Traditionally, the strength values were based on the failure strength of postmortem pine segments under axial strength values were based on the failure strength of postmortem spine segments under axial load. Lately, spine segments were tested under cyclic load, and fatigue failure data of spine segments are made available (Brinckmann *et al.* 1987). Often a computerized biomechanical model of the human musculoskeletal system is used to predict human capabilities for a particular task performance (Chaffin 1997). To predict whole-body exertion capabilities for a given population, Chaffin (1997) presents example logic for a model (Figure 7.4). Specific muscle group strength data and spinal vertebrae failure data are used in this model as the limiting values for the reactive movement at various body joints when an operator of particular stature and body weight attempts an exertion, such as lifts, pushes or pulls in a specific direction with one or both hands while maintaining a predicted stature. For comparing task exertions where specific localized muscle actions exist, it is possible to use electromyography (EMG) for the assessment of the active muscles and/or a subjective discomfort rating method.

Several guideline and design evaluation methods are available for the initial design of a workstation. The selection of an appropriate method will depend of the objective in terms of solving a specific problem, and whether a single exertion or multiple exertions are involved (Figure 7.5; Chaffin 1997). A prototype workstation design can be evaluated by employing a suitable method or several methods presented in Figure 7.6; Chaffin 1997). A representative worker can be used to further refine the initial design.

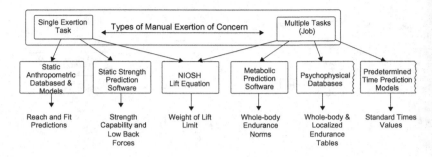

FIGURE 7.5   Models and databases available for proscriptive guidance in design of new work-places with outcome predictions listed at bottom of each model or database (Chaffin 1997).

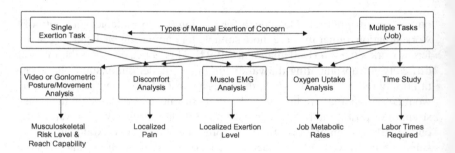

FIGURE 7.6   Methodologies for measuring human performance during prototype analyses, with outcome data listed for each (Chaffin 1997).

7.5.6 Workstation Design for Manual Materials Handling

From a biomechanical viewpoint, package size can be of problem, especially when the package is located on or near the floor (Chaffin 1997). If the load is of a size that cannot be easily handled between the knees at the start of the lift, then an operator must lean the torso forward. Using the 3D Static Strength Prediction Program, Chaffin (1994) had demonstrated the effect of lifting two different size boxes of the same weight. The combination of a forward torso angle and large horizontal distance between the large box and low back caused about a 30–38% increase in predicted L5/S1 disc compression force compared to small box lifting close to the body. A significant risk of low back injury would result, if the object weighed >35 lb.

From the viewpoint of workstation design, when large packages are involved they should never be presented to an operator at a height lower than about mid-thigh (or ~30 inches). This would allow the operator to stand erect and bring the object against (or near) the torso, thus minimizing lower back bending movements and resulting spinal compression forces. When large bulky objects are involved, adjustable lift tables should be used (Chaffin 1997).

The orientation of the package when presented to the operator must be taken into consideration. If the shape of the package is not like a cube but rather one small dimension, it should be presented in a more vertical direction. This will permit the operator to list the object closer to the body by straddling the narrow dimension, or if a handle is provided on the top, by lifting close to the side of the body.

7.6 SUMMARY AND CONCLUSIONS

For the design of an ergonomic workstation, it is essential to give utmost consideration to engineering anthropometry and occupational biomechanics. This will ensure a reduction of many risk factors of occupational injury and contribute toward enhancing worker productivity, worker physical and mental well-being and job satisfaction.

In a workstation design, an ergonomic analysis is performed to deal adequately with spatial accommodation, posture, reaching abilities, clearance and interference of the body segments, field of vision, available strength of the operator, and biomechanical stress. Typically used in the analysis are the appropriate anthropometric masses and inertial properties from the established databases. It is necessary to make anthropometric data adjustment to account for clothing, shoe and slump posture. A systematic ergonomics approach is provided for the determination of workstation design parameters. The four essential design dimensions are: (1) work height, (2) normal and maximum reaches, (3) lateral clearance and (4) angle of vision and eye height. A case study of a supermarket checkstand workstation is provided to illustrate the systematic manner by which design parameters can be determined. To evaluate human–machine interaction for the design of a workstation, the use of computerized human modeling programs is illustrated by presenting two representative models: JACK and MANNEQUIN. The recent developments in workstation design include: (1) a computerized potentiometric system for structural and functional anthropometric measurements and (2) an improved anthropometric model of 3D maximum reach envelope.

A biomechanical evaluation of a workstation is necessary, when a worker is believed to be exposed to excessive physical stress or risk in a workstation. It is possible to minimize or eliminate work-related injuries to musculoskeletal system through redesign of a workstation. For the determination of a person's physical capabilities, dynamic strengths measurement is more appropriate than isometric or static strengths measurement. Guidelines are provided so that industrial workers do not overload muscles to avoid or minimize fatigue. It is important to determine human strength profiles in the workspace for optimum design of a workstation. Since back injury often causes low back pain, it is necessary to determine the stresses at lower back due to work performed at different trunk postures. The compressive force generated at lower back (L5/S1) is believed to be the most significant cause for low back pain. To predict human capabilities for a particular task performance, a computerized biomechanical model of the human musculoskeletal system is often used. For the initial design of a workstation, several guideline and design evaluation methods are available. The effect of lifting two different size boxes of the same weight was demonstrated by using a 3D Static Strength Prediction Program. Guidelines are provided for manual materials handling especially when large packages are involved.

REFERENCES

ANDERSSON, G., ORTENGREN, R. and HERBERT, P., 1977, Quantitative electromyographic studies of back muscle activity related to posture and loading. *Orthopedic Clinics of North America*, 8, 85–96.

ANDERSSON, G.B.J., POPE, M.H., FRYMOYER, J.W. and SNOOK, S., 1991, Epidemiology and costs. In Pope, M., Andersson, G.B.J., Frymore, J. and Chaffin, D.B. (eds) *Occupational Low Back Pain* (St Louis: Mosby Year Book), pp. 95–113.

AYOUB, M.M. 1973, Workplace design and posture. *Human Factors*, 15, 265–268.

AYOUB, M.M. and MITAL, A. 1989, *Manual Materials Handling* (London: Taylor & Francis).

BADLER, N.I. 1991, Human Factors Simulation Research at the University of Pennsylvania. Computer Graphic Research Quarterly Progress Report No. 38, Department of Computer and Information Science, University of Pennsylvania, Fourth Quarter 1990, 1–17.

BRINCKMANN, P., JOHANNLEWELING, N., KILWEG, D. and BIGGEMANN, M., 1987, Fatigue fracture of human lumber vertebrae. *Clinical Biomechanics*, 2, 94–96.

BURDORF, A., GOVAERT, G. and ELDERS, L. 1991, Postural load and back pain of workers in manufacturing of prefabricated concrete elements. *Ergonomics*, 34, 909–918.

CHAFFIN, D.B., 1973, Localized muscle fatigue — definition and measurement. *Journal of Occupational Medicine*, 15, 346–354.

CHAFFIN, D.B., 1994, Postural considerations in lifting — or why aren't my arms five feet long? In Nordin, M., Andersson, G. and Pope, M. (eds) *Occupational Musculoskeletal Disorders: Assessment, Treatment and Prevention* (St Louis: Mosby Year Book).

CHAFFIN, D.B., 1997, Biomechanical aspects of workplace design. In Salvendy, G. (ed) *Handbook of Human Factors and Ergonomics* (2nd ed.) (New York: Wiley), pp. 772–789.

CHAFFIN, D.B. and ANDERSSON, G.B.J., 1984, *Occupational Biomechanics* (New York: Wiley).

CHAFFIN, D.B. and ANDERSSON, G.B.J., 1991, *Occupational Biomechanics* (2nd ed.) (New York: Wiley).

CHAFFIN, D.B. and PARK, K.S., 1973, A longitudinal study of lowback pain as associated with occupational weight lifting factors. *American Industrial Hygiene Association Journal*, 34, 513–525.

DAS, B. and BEHARA, D.N., 1995, Determination of the normal horizontal working area: A new model and method. *Ergonomics*, 38, 734–748.

DAS, B. and BLACK, N.L., 2000, Isometric pull and push strengths of paraplegics in the workspace: 1. Strength measurement profiles. *International Journal of Occupational Safety and Ergonomics*, 6(1), 47–65.

DAS, B. and FORDE, M., 1999, Isometric push-up and pull-down strengths of paraplegics in the workspace: 1. Strength measurement profiles. *Journal of Occupational Rehabilitation*, 9, 279–291.

DAS, B. and GRADY, R.M., 1983a, Industrial workplace layout design: an application of engineering anthropometry. *Ergonomics*, 26, 433–447.

DAS, B. and GRADY, R.M., 1983b, The normal working area in the horizontal plane: A comparative analysis between Farley's and Squires' concepts. *Ergonomics*, 26, 449–459.

DAS, B., KOZEY, J.W. and TYSON, J.N., 1994, A computerized potentiometric system for structural and functional anthropometric measurements. *Ergonomics*, 37, 1031–1045.

DAS, B. and SENGUPTA, A.K., 1995, Computer-aided human modelling programs for workstation design. *Ergonomics*, 38, 1958–1972.

DAS, B. and SENGUPTA, A.K., 1996, Industrial workstation design: a systematic ergonomics approach. *Applied Ergonomics*, 27, 157–163.

DAS, B. and WANG, Y., 1995, Determination of isometric pull and push strength profiles in workspace reach envelopes. In Proceedings of the Annual International Industrial Ergonomics and Safety Conference, Seattle, WA, pp. 1017–1024.

DAVIS, P.R. and STUBBS, D.A., 1977, Safe levels of manual forces for young males (2). *Applied Ergonomics*, 8, 141–150.

DELLEMAN, N.J., DROST, M.R. and HUSON, A., 1992, Value of biomechanical macromodels as suitable tools for the prevention of work-related low back problems. *Clinical Biomechanics*, 7, 38–48.

FARLEY, R.R., 1955, Some principles of methods and motion study as used in development work. *General Motors Engineering Journal*, 2, 20–25.

GENAIDY, A.M., AL-SHEDI, A. and KARWOWSKI, W., 1994, Postural stress analysis in industry. *Applied Ergonomics*, 25, 77–87.

GRANDJEAN, E., 1982, *Fitting the Task to the Man: An Ergonomic Approach* (3rd ed.) (London: Taylor & Francis).

HERTZBERG, H.T.E., 1972, Engineering ~~and anthropometry~~. In van Cott, H.P. and Kinkade, R.G. (eds) *Human Engineering Guide to Equipment Design* (New York: McGraw-Hill), pp. 467–584.

HUMANCAD, 1991, *MANNEQUIN User Guide* (New York: Melville).

HUNSICKER, P., 1955, Arm Strength at Selected Degrees of Elbow Flexion. WADC Technical Report 54–548. United States Air Force, Project 7214.

JUUL-KRISTENSEN, B., FALLENTIN, N. and EKDAHL, C., 1997, Criteria for classification of posture in repetitive work by observation methods: a review. *International Journal of Industrial Ergonomics*, 19, 397–411.

HEINSALMI, P., 1986, Method to measure working posture loads at working sites (OWAS). In Corlett, E.N., Wilson, J. and Manenica, I. (eds) *The Ergonomics of Working Postures. Models, Methods and Cases* (London: Taylor & Francis), pp. 100–104.

KARHU, O., KANSI, P. and KOURINKA, I., 1977, Correcting working postures in industry: A practical method for analysis. *Applied Ergonomics*, 8, 199–201.

KARWOWSKI, W. 1992, Occupational biomechanics. In Salvendy, G. (ed) *Handbook of Industrial Engineering* (2nd ed.) (New York: Wiley), pp. 1005–1046.

KROEMER, K., KROEMER, H. and KROEMER-ELBERT 1996, *Ergonomics: How to Design for Ease and Efficiency* (Englewood Cliffs: Prentice Hall), pp. 66–67.

KUMAR, S., 1991, Arm lift strength in workspace. *Applied Ergonomics*, 22, 317–28.

MARRAS, W.S., 1997, Biomechanics of the human body. In Salvendy, G. (ed) *Handbook of Human Factors and Ergonomics* (2nd ed.) (New York: Wiley), pp. 233–267.

MITAL, A. and DAS, B., 1987, Human strength and occupational safety. *Clinical Biomechanics*, 2, 97–106.

MITAL, A. and FAARD, H., 1990, Effects of sitting and standing, reach distance and arm orientation on isokinetic pull strengths in the horizontal plane. *International Journal of Industrial Ergonomics*, 6, 241–248.

MITAL, A., KARWOWSKI, W., MAZOUZ, A.K. and ORSARH, E., 1986, Prediction of maximum acceptable weight of lift in the horizontal and vertical planes using simulated job dynamic strengths. *American Industrial Hygiene Association Journal*, 47, 288–292.

PINZKE, S., 1992, A computerized method of observation used to demonstrate injurious work operations. In Mattila, M. and Karwowski, W. (eds) *Computer Applications in Ergonomics, Safety and Health* (Amsterdam: Elsevier/North-Holland). pp. 359–364

PHEASANT, S., 1986, *Body Space: Anthropometry, Ergonomics and Design* (London: Taylor & Francis).

PUTZ-ANDERSON, V., 1994, *Cumulative Trauma Disorders: A Manual for Musculoskeletal Diseases of the Upper Limbs* (London: Taylor & Francis).

SENGUPTA, A.K. and DAS, B., 1998, A model of three-dimensional maximum reach envelope based on structural anthropometric measurements. In Proceeding of the Annual International Occupational Ergonomics and Safety Conference, Ypsilanti, MI, pp. 226–229.

SENGUPTA, A.K. and DAS, B., 1999, The determination of maximum reach envelope for industrial workstation design. In Proceedings of the Annual International Society for Occupational Ergonomics and Safety Conference, Orlando, FL, pp. 59–64.

SNOOK, S.H. and CIRIELLO, V.M., 1991, The design of manual material handling tasks: revised tables of maximum acceptable weights and forces. *Ergonomics*, 34, 1197–1213.

SQUIRES, P., 1956, *The Shape of the Normal Work Area*. Report no. 275 (New London: Navy Department, Medical Research Laboratory, CT).

TICHAUER, E.R., 1975, *Occupational Biomechanics: The Anatomical Base of Workplace of Workplace Design*. Monograph No. 51 (New York: Institute of Rehabilitation Medicine, New York University, Medical Center Rehabilitation).

WATERS, T.R. and PUTZ-ANDERSON, V., 1996, Manual material handling. In Bhattacharya, A. and McGlothlin, J.D. (eds) *Occupational Ergonomics: Theory and Applications* (New York: Marcel Dekker), pp. 329–349.

FURTHER READING

Abd Rahman, N. I., Md Dawal, S. Z., Yusoff, N., & Mohd Kamil, N. S. (2018). Anthropometric measurements among four Asian countries in designing sitting and standing workstations. *Sādhanā, 43*, 1–9.

Colim, A., Carneiro, P., Costa, N., Arezes, P. M., & Sousa, N. (2019). Ergonomic assessment and workstation design in a furniture manufacturing industry—a case study. In: Arezes, P., et al., *Occupational and Environmental Safety and Health* (pp. 409–417). Cham: Springer International Publishing.

Kushwaha, D. K., & Kane, P. V. (2016). Ergonomic assessment and workstation design of shipping crane cabin in steel industry. *International Journal of Industrial Ergonomics, 52*, 29–39.

Sutalaksana, I. Z., & Widyanti, A. (2016). Anthropometry approach in workplace redesign in Indonesian Sundanese roof tile industries. *International Journal of Industrial Ergonomics, 53*, 299–305.

Woo, E. H. C., White, P., & Lai, C. W. K. (2016). Ergonomics standards and guidelines for computer workstation design and the impact on users' health–a review. *Ergonomics, 59*(3), 464–475.

Anthropometric Topography

8

Z. Li

8.1 INTRODUCTION

The term "anthropometric topography" can be defined as the measurement and study of the configuration of human body surface, including its relief, features and structural relationships. The main differences between anthropometric topography and traditional anthropometry include: (1) Anthropometric topography emphasizes the measurement of body surface rather than pairs of landmarks (points); (2) features including not only dimensions are extracted from body surface measurement; (3) among the features, structural relationships, but not just value relationships, are a very important outcomes of the study, while traditional anthropometric measurement generally loses the information of structural relationships of dimensions. Thus, anthropometric topography can provide more useful information concerning the human body for ergonomic considerations in product design, occupational safety and health evaluation, human modeling and simulation, and so on. Among the content of anthropometric topography, dimensions, shape, and spatial distributions are mostly concerned with the context of ergonomics.

Extensive work has been carried out on human body surface measurement using contemporary three-dimensional (3D) scanning technologies, which is called 3D (surface) anthropometry. Mass discrete points captured during the scanning form a "point cloud" representing the surface of the body. Polygonal surface rendering is often used to visualize the point cloud to a 3D shape. The result of 3D scanning, i.e. relief of human body surface, can be directly used in

DOI: 10.1201/9781003459767-8

customized product design and evaluation. However, further data processing, which can be very complicated, is required for population measurement and study. Currently, the major work of data processing focuses on landmark labeling and extraction of human body dimensions from 3D scanning—it can be treated as a traditional anthropometric survey using 3D scanning technologies rather than anthropometers, calipers, tapes, etc. Sizing and grouping based on human body surface shape and spatial distribution are more prospective for better population fitting. To do this, advanced features of human body surface need to be defined and extracted from 3D anthropometric data. Indexing and retrieving are often necessary for the database management of 3D human body surface topography. Extracted features are also needed. Other data processing work includes hole-filling, data filtering, surface modeling, registration and alignment of multiple scans, and so on. For various application purposes, special data processing may be required. For example, segmentation of human body scans is crucial to lifelike human modeling and animation.

8.2 THREE-DIMENSIONAL ANTHROPOMETRIC TECHNOLOGIES

There were several human body surface capturing methods even before contemporary 3D scanning technologies became available. These methods may be crude, but they are practical. Figure 8.1 shows a simple device used in China to measure body shape for garment model development. For example, when measuring one

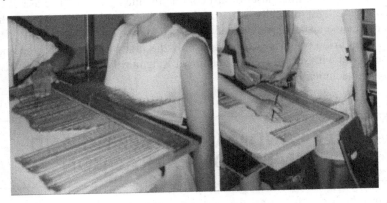

FIGURE 8.1 Device used in China to measure body shape for garment model development.

section of the front of the trunk, sliding sticks constrained by a frame can be moved to touch body surface. Body shape is then mapped to end points of the sticks based on which a section curve can be drawn. The measurement is conducted section by section. A front curve and a back curve in a same section can be merged together according to dimensions such as thickness in the section.

When photograph technologies emerged, these were used to capture human body silhouettes. Photos taken from arranged directions can be used to reconstruct body bulge. However, since image size is related to object distance in a lens-imaging system, photos of 3D human body shape is inherently distorted. It can be easily verified, by experiments or calculation, that measurement of dimensions at different distances from the camera, using a single scale from one photo, often introduces unacceptable errors for anthropometric survey.

Besides reverse engineering for product design, clinical application, especially reconstructive surgery, is another major driving force of the development of 3D surface topography measurement technologies. Beginning in 1980, a series of international symposiums on human body surface topography and spinal deformity provided rich information regarding both technologies and applications.

Currently, structured light scanning has become a dominant technology in surface capturing due to advantages such as noncontact measurement, highspeed, high density of point cloud, etc. In this method, structured (coded) light is projected onto the surface to be measured and is then distorted by the modulation of surface relief. The distortion is captured, typically by two cameras (one camera systems are already available). Based on phase-shift, triangular intersection, and other related theories, a central processing unit or computer then analyzes the images and determines 3D coordinates of surface points. The light can be lined laser or just common white light grid. The use of common white light normally brings higher efficiency and is safer. Major wholebody scanners take approximately 5–20 sec for one scan.

Capturing of mass point data is only the first step in 3D anthropometry. Compared with scanning, post data processing may be far more time consuming and technically complicated.

8.3 LANDMARK LABELING AND DIMENSION EXTRACTION

Human body dimensions are the most common features of human body topography although they were normally not considered as matter of surface topography. Only after 3D anthropometry technologies emerged were dimensions linked with surface topography—they are extracted from surface topographic data rather than directly measured. Dimensions can be treated as macro-level features. Necessary software is required to help label/identify anthropometric landmarks and calculate dimensions based on anthropometric definitions. Although providers of whole body scanners may declare that their accompanied software can automatically label most anthropometric landmarks, it should be pointed out that it is actually difficult or error-prone to identify anthropometric landmarks even on textured body surface scans, since many of the anthropometric landmarks are defined according to bones which are invisible. Checking and relocating landmarks may be time-consuming and the operator will face inevitable uncertainty. Thus, landmarks should be annotated on subject body surface before scanning. There are several notating methods, e.g. small round stickers, cosmetic labeling, and so on. The color of landmarks should be compatible with the scanning technology so that they are distinctive in the resulting scan.

In practice, the new anthropometric method has no remarkable advantages regarding dimensions. Instead, its disadvantages are obvious: expensive investigation, complicated operation, inconvenience in moving/reinstalling, timecosting post data processing, mass memory storage, and so on. The most important reason for using the new technology is the capturing of body surface topography. Apart from customized design, the benefits of body topographic information are being developed. Even the study of spatial relationship of landmarks can provide potentially improved population grouping, as has been investigated by Mochimaru *et al.* (2000).

8.4 SHAPE DESCRIPTION AND COMPARISON

Shape is a very important factor regarding anthropometric topography. From a very earlier age, e.g. ancient Greek, attempts have been made to characterize the shape of the human body. The well-known Sheldon somatotype theory was developed in the 1940s based on body measurements and photographs. Three basic body types (i.e. somatotypes) were identified: endomorph, mesomorph, and ectomorph. There are many other shape classifications of human body surface and organs in anthropology, cosmetology, and restoration studies.

Features to describe human body shape, called shape descriptors, need to be established before characterizing human body shape. Perhaps the simplest descriptor of human body shape that has been used is the ratio of stature and weight. In the clothing industry, waist, chest, or hip circumferences may be considered, besides stature and weight.

Volume and mass may also be used to differentiate objects, but they are not shape features. In biological morphology studies, inertia and distance from axes are often used to characterize similarity of life forms that are aligned by center of mass and principle inertial axes. The method can be used in indexing and retrieving human body scan databases, such as Paquet *et al.* (2000), but the descriptors do not directly characterize shape.

When spatial relationship is to be considered, it is natural to establish shape descriptors based on landmarks. FiniteElement Scaling Analysis, Procrustes Analysis, Euclidean Distance Matrix Analysis, and Thin Plate Splines use landmarks to describe and compare shapes.

Mathematical models in the field of computer graphics describe shape mostly in an implicit manner. B-spline, Non-Uniform Rational B-Splines (NURBS), and other models in surface modeling, can surely be used in describing local human body surfaces. For complex surfaces such as the hand, the polygonal surface model is more suitable. Shapes can then be compared based on surface model parameters and geometrical properties such as radius. Reverse engineering technologies can be helpful in constructing a geometrical model from a raw scan of human body surface.

The Fourier method can analyze shape in a more complicated and in-depth way. It is often used in signal and optical analysis, afterwards adopted in 2D shape analysis, and has been used in 3D shape analysis from approximately 1990. It transforms 3D shape into sines and cosines functions, and coefficients of the functions, rather than point coordinates are used in further analysis. However, there is minimal literature on using the Fourier method in 3D shape analysis for sizing and population grouping. In the thin plate splines method, a special descriptor, energy to transform a thin plate to a target shape, is used to characterize shape. The method can be used to compare two shapes with different numbers of surface points.

Wavelet transformation, another useful tool for image processing and signal analysis, is a prospective method for shape analysis and comparison. By transformation, a surface can be decomposed into multiresolution descriptions. The low-resolution part describes the basic shape of a surface with a small number of control points, while the high-resolution part describes details of the surface. The low-resolution part is useful for shape comparison and classification. Ignoring details of human body surface can avoid the influence of measurement noise and micro relief such as wrinkles. The resolution of decomposition can be selected to gain a proper approximation of the original surface. Here, coefficients of wavelets are good shape descriptors for shape analysis and comparison. As in the Fourier method, wavelet transformation has not been successfully applied in industry because of its complex nature.

8.5 INDEXING AND RETRIEVING

In traditional dimension measurement, raw data are onedimensional data that can be processed by the classical statistical method. The survey results, such as percentiles, can be published in tables. Indexing and retrieving of the databases can be based only on existing data items. However, the matter is very different to anthropometric topographic survey because the raw data are 3D point data. Statistical methods to analyze topographic sample data are discussed in the following section. For indexing and retrieving of sample databases, extracted dimensions can be used, but other extended descriptors can also be helpful. Extended descriptors are normally established as keys. The descriptors can be just text description and/or keywords of the sample, codes given according to characteristics of samples, or descriptors of point cloud structure/distribution and surface shape. For example, moment of inertia, average distance of surface from the axis of inertia, and variance of distance of the surface from the axis are employed in 3D shape similarity analysis for retrieving models in some studies. In these studies, principal axes of inertia are often used in pose normalization.

8.6 SIZING AND POPULATION GROUPING

8.6.1 Introduction

For better fit with human body surface, products like some gauze masks are designed with components that can be deformed by hand to adapt to human body surface. However, not all products can be designed in this way, otherwise their functions cannot be satisfied. For instance, mechanical energy protectors require strong structures. Some products, such as goggles, may be designed with soft interfaces with human body surfaces. In population accommodation, this solution is not good enough if the sizing system is not elaborated with considerations of human body shape as uneven pressure applied by soft components to human body surface can be intolerably uncomfortable, especially when worn for long durations.

A potential significant benefit of 3D anthropometric topography study is sizing and population grouping, based on human body shape, for better fitting design for products which interact with body surface when they are worn or function, such as face shields, glasses and eye protectors, earphones, helmets, masks, clothes, shoes, paraphernalia, hand-held tools, and so on.

8.6.2 Traditional Sizing and Grouping Method

Traditional sizing and grouping is based on the analysis of principle dimensions/parameters that account for as much of the variability and thus effectively differentiate the population on the context of intended applications. Statistical methods such as Principal Component Analysis (PCA) can be helpful in identifying principle dimensions. In simple cases, one dimension or combined parameter (such as ratio of stature and weight) may be enough to separate samples into groups with satisfying quality for the applications. More frequently, bi-variate analysis is used where two principle dimensions are identified for sizing and grouping. For example, hand-length and hand-breadth at metacarpals were used in Chinese adult hand sizing (GB/T 16252-1996). In a subgroup, appropriate relations between other dimensions and principle dimensions can be established by regression analysis so that they can be conveniently calculated from principle dimensions when designing related products. Multi-variate analysis is more complicated and rarely used, although it may result in a more aborative sizing scheme.

Unfortunately, the statistical method traditionally used in anthropometry survey cannot be applied directly to 3D anthropometric data (point coordinates); otherwise it could lead to misinterpretations. Taking percentile concept as an example, it is defined strictly for estimates of onedimensional anthropometric data. With more design parameters (e.g. dimensions) considered with a same percentile, less population will be covered, since less people can be the percentile for all the parameters. However, in many if not most anthropometric and biomechanical analysis software, predefined human models often have identical percentiles. Thus, inherently there are mistakes when using these models in design of workstations, machines and so on (explained in Section 7). At the least, somatotypes should be considered in the models to reduce mistakes. Three-dimensional geometrical human models have similar problems. Another error-leading problem is statistics of dimensions based on coordinates. It is mathematically shown that dimension percentile calculation based on coordinate statistics is incompatible with those traditionally based on individual dimensions (Cai *et al.* 2005). Similarly, if statistical methods used in sizing and grouping are applied directly to 3D anthropometric data, there are potential mistakes. Thus, methods for sizing and grouping based on shape are quite different from traditional methods.

8.6.3 Sizing and Grouping Based on Shape

Sizing and population grouping based on shape is a very complicated problem that has not been adequately studied and practiced. Shape concerns not only 3D constructing or control points, but also their internal structure. Feature extraction and comparison should consider both issues. A generalized distance concept is usually developed based on shape descriptors (discussed in Section 4) to characterize the difference of shape between a surface sample and a common reference, which can be one of the surface samples or a predefined standard shape. In distance calculation, a surface is represented by selected landmarks or other sampled surface points (sampling method should be selected firstly), control points in surface mathematical model, and other extended surface descriptors. Mochimaru *et al.* (2000) proposed a homologous shape modeling method to design well-fitting products, in which a part of the body is represented by the same number of data points of the same topology, and each data point is defined based on landmarks. The Free Form Deformation (FFD) technique was used to analyze human body forms and to classify them into several groups. The FFD method deforms the shapes of objects smoothly by moving control lattice points set around the object. The reference body form is automatically deformed to coincide with the other body forms using the FFD method. Dissimilarity was identified by movements of control lattice points. The method has been used to analyze foot forms and faces for sizing. After grouping, a representative shape was selected for each group.

In some cases, sizing/grouping should be considered based on the spatial distribution of interested points/zones with reference of surface. For example, Whitestone and Robinette (1997) provided a solution to the design of a helmet integrated with displays that needed to consider spatial distribution of the pupils. A critical factor, the relationship of the head to the helmet, is emphasized to define anthropometric envelopes with discussion of various design myths.

They pointed out a fundamental fact that is often neglected—helmet systems do not fit the human head in exactly the same way across a sample of people. By 3D surface scanning of the head with a prototype helmet system, and the prototype only, the spatial relation of the head and the helmet system in function can be obtained. Then, all head and face data are referenced to a common helmet-based coordinate system. After that, design envelopes can be derived under the condition that the helmet fits the head. Finally, viewing devices or acoustic equipment can be designed. Here, the helmet surface acts as a reference surface.

8.6.4 Boundary Issue

In sizing of shape, the boundary of interested body surface area is sometimes an important factor and should be defined according to functional requirements on either body surface or product surface. From the point of view of target users of a product, surfaces on their body to be accommodated are diverse, not only on surface shape, but also surface scope. For example, a design is required for a protective helmet to protect the top of the head and the ears from flying objects and noise. In this case, the surface boundary should be a little beyond the ears and is quite diverse from one person to another. In more cases, boundary can be considered from the point of view of the product—a definite interfacing area corresponds to different functional areas of users. In these cases, the surface boundary of the human body is determined according to a predefined boundary of interfacing surface of the product.

8.6.5 Normalization and Scaling Issue

Normalization or scaling is often a preprocessing procedure to make all samples have the same value of variables, except the ones applied to the selected analysis method. However, normalization or scaling may change original object shape. In Mochimaru's method, all samples were normalized to have a same value of foot length before further analysis. This procedure actually changes the shape and may introduce potential errors: Two samples with very different sizes may be normalized to have the same shape and classified into the same group.

8.6.6 Alignment Issue

Another important issue to be considered before sizing and grouping analysis is the alignment of samples. Some artificially defined references, such as the Frankfurt plane, have been used to align samples. Whitestone et al. (1997) pointed out possible problems in using these references. In their method, head and face surfaces were aligned actually by referencing to another object, a helmet, but not corresponding points or surfaces of the samples. Cai et al. (2005) analyzed the influence of alignment references on 3D anthropometric statistics mathematically. It was concluded that different alignment reference points (e.g. landmarks) for translation alignment could result in different object shapes if 3D anthropometric data are processed for percentile values based on coordinates (as mentioned above, this is misleading). The analysis implies that the alignment reference points for 3D anthropometric statistics should be carefully chosen. If statistics based on coordinates have to be calculated, the alignment reference points should be selected depending on the particular needs for each application. Also, the validity of the results is limited to the particular application and will not be compatible with traditional dimensional statistics. Alignment by artificial references is subjective and ambiguous. Shape classification will strictly correspond to its application purpose. Thus, functional points and surfaces should be considered as alignment references in general. To align two 3D shapes, six freedoms need to be fixed. Although references corresponding to some freedoms are normally easy to determine, other references are not. Other objects, such as the helmet mentioned above, may be necessary for alignment referring.

8.7 HUMAN MODELING, ANIMATION, AND SIMULATION

Many human modeling, animation, or simulation systems have been developed for industrial application or academic research, some of which are commercially available. Some systems for motion or biomechanical analysis may use stick figures, concerning dimensions of an individual subject. More systems try to use 3D mannequins with statistical dimensions and representative body shape. The most common mannequins include the P5, P50, and P95 models for both males and females. It should be noted that these identical percentile models actually represent nobody—body dimensions are not proportional. For many applications, using these ideal models is acceptable. However, sometimes it provides misleading results. One reason for the model simplification lies in the absence of 3D "statistical human models." Representative models of different somatotypes of populations may solve the problem to some extent, but extensive research and survey work is still required.

8.8 CONCLUSIONS

Anthropometric topography can be regarded as a revolutionary development of traditional anthropometry under the background of emergence of various 3D measurement technologies, although it is still in an early stage of development. In this chapter, the concept of anthropometric topography is introduced. 3D anthropometric measurement technologies are discussed. Dimensions are regarded as macro-level features of human body surface. Extraction of dimensions from 3D anthropometric topography is discussed in Section 3. As an important feature of human body surface, shape is discussed in more detail, including its description, indexing, and retrieving, and especially its comparison and grouping. Finally, some aspects of human modeling, animation, and simulation concerning anthropometric topography are discussed.

ACKNOWLEDGMENT

The author is grateful to Mr. Hui Xiao, senior engineer of China National Institute of Standardization, for providing the photograph of shape measurement for garment model development.

REFERENCES

CAI, X.W., LI, Z.Z., CHANG, C.C. and DEMPSEY, P., 2005, Analysis of alignment influence on 3D anthropometric statistics. *Tsinghua Science and Technology*, 10(5), 623–626.

MOCHIMARU, M., KOUCHI, M. and DOHI, M., 2000, Analysis of 3-D human foot forms using the free form deformation method and its application in grading shoe lasts. *Ergonomics*, 43(9), 1301–1313.

PAQUET, E., RIOUX, M., MURCHING, A., NAVEEN, T. and TABATABAI, A., 2000, Description of shape information for 2-D and 3-D objects. *Signal Processing: Image Communication*, **16(1), 103–122.**

WHITESTONE, J.J. and ROBINETTE, K.M., 1997, Fitting to maximize performance of HMD systems. In Melzer, J.E. and Moffitt, K.W. (eds) *Head-Mounted Displays: Designing for the User* **(New York: McGraw-Hill), pp. 175–206.**

FURTHER READING

Hojaij, F. C., Docko, A., Franceschi, L., Yendo, T. M., Akamatsu, F., Jacomo, A. L., & Cernea, C. R. (2019). Thyroid gland topography. *Archives of Head and Neck Surgery*, *47*(4), 0–0.

Karl H. E. Kroemer, Hiltrud J. Kroemer, Katrin E. Kroemer-Elbert (eds.) (2020). Engineering Anthropometry. *Engineering Physiology: Bases of Human Factors Engineering/Ergonomics*, 299–364.

Kumar, A., Kumar, A., Sinha, C., Sawhney, C., Kumar, R., & Bhoi, D. (2018). Topographic sonoanatomy of infraclavicular brachial plexus: Variability and correlation with anthropometry. *Anesthesia, Essays and Researches*, *12*(4), 814.

Prowse, A., Pope, R., Gerdhem, P., & Abbott, A. (2016). Reliability and validity of inexpensive and easily administered anthropometric clinical evaluation methods of postural asymmetry measurement in adolescent idiopathic scoliosis: a systematic review. *European Spine Journal*, *25*, 450–466.

Shan, Z., Hsung, R. T. C., Zhang, C., Ji, J., Choi, W. S., Wang, W., Yang, Y., Gu, M., & Khambay, B. S. (2021). Anthropometric accuracy of three-dimensional average faces compared to conventional facial measurements. *Scientific Reports*, *11*(1), 12254.

Anthropometry of Children

9

E. Nowak

9.1 INTRODUCTION

When we wish to adjust environments to meet human needs, it is necessary to take into account the age of the persons using them. In all environments, the anthropometric data of children and juveniles play an important role in the design of toys, interior furnishings and furniture, kindergartens, schools, hospitals, playgrounds, etc.

A young and not completely developed organism is extremely susceptible to the influence of the external environment. Furniture, equipment and toys, if wrongly designed and not adjusted to the characteristics of a young child, can result in defective body posture and the establishment of pathological states. In order to prevent this, it is necessary to use various remedial measures. As well as providing children with proper medical care and correct nourishment, it is necessary for their development to consciously shape the surroundings, including the world of objects. Correct design, taking note of ergonomic criteria, can facilitate good physical development in a child, while the use of harmonious forms and rational solutions helps to form aesthetic feelings and improve the quality of everyday life. Ergonomic values should be apparent in every object intended for use by the youngest generation.

Anthropometry is one important element that has an effect upon the ergonomics of the child's environment. It provides data concerning the physical development of children and the changes that take place in body proportions with age. It also provides information about the intergeneration differences in

DOI: 10.1201/9781003459767-9

somatic features observed in a given population. This is the so-called phenomenon of secular trend. When this phenomenon occurs in a given population, a designer should use current anthropometric data only. In this respect, the Institute of Industrial Design, among others, specializes in the environmental ergonomics of children and young people, and conducts research to develop anthropometric data characterizing the basic design requirements of children.

This article describes the successive phases of the investigation involving Polish children and presents their actual somatic characteristics against the development of European children. Application of anthropometric data to the design of school furniture and development of standards is also discussed. The article also introduces two-dimensional manikins which illustrate, to scale, the changing physical shape of a child's body with age, and are of assistance in designing objects meant for children and the young.

9.2 CHARACTERISTICS OF THE SOMATIC DEVELOPMENT OF CHILDREN AND JUVENILES: DATA FOR DESIGNING

Before a proportionally shaped adult figure is finally formed, it undergoes many intermediate states, which are characterized by different body proportions. This results from the irregular development of the body systems, such as the skeletal and muscular systems, in particular developmental periods, and different growth rates of individual parts of the body. In relation to adults, newborns and, to a lesser degree, infants, have large heads, short necks, long trunks and short limbs, especially lower limbs. In the course of development, from the moment of birth to adulthood, the trunk, already proportionately long in infants, increases to three times its original length. At the same time, muscle mass also increases. It is estimated that the muscular system increases by as much as 42 times its original size.

Apart from this basic information concerning changes in the structure of the human body, designers need to become acquainted with some of the biological phenomena that are directly connected with physical development. These are the secular trend phenomenon, the acceleration process and the pubertal spurt of body growth.

The secular trend is defined as the sequence of changes of human physical features, which occur gradually from one generation to another, tending in a constant direction. These changes mainly concern stature dimensions—e.g. children are taller than their parents and have bigger body size.

Acceleration of development is understood to be the inter-generational speeding-up of biological development and puberty, and thus the earlier arrival of consecutive phases of development. The acceleration of development of the whole organism in every successive generation by comparison to the preceding one initially concerns puberty. One of the manifestations of this period is the so-called pubertal spurt, its characteristic being that the yearly increase in height just before the onset of puberty is greater than in previous years in both boys and girls.

Differences in body dimensions of adult men and women result, to a large extent, from the differences occurring in the course of the pubertal spurt. Before this period, the young of both sexes have similar body height. Since the pubertal spurt of girls begins earlier, it is less intensive; the process of body growth also stops earlier; this is why adult women have significantly lower stature values. In general, it is estimated that comparable body dimensions of women amount to about 90–96% of the average dimensions of men (Flügel et al. 1986). The difference in stature amounts to 8–10 cm. On average, the body mass of women is 8kg lower than that of men. Those disproportions will be bigger when taking into account extreme percentile values.

Investigations concerning the secular trend have been undertaken by many research centers all over the world.

Reflections on the reasons which determine the strength and course of secular trend are very interesting. The majority of scientists are very careful, however, about drawing final conclusions. Numerous data suggest that in the 19th century, secular trend was inconsiderable; its onset was connected with the industrial revolution. In general, scientists agree that environmental conditions are of significant importance in the observed phenomenon of secular trend. Economic prosperity is conducive to the occurrence of this phenomenon—living conditions and nourishment improve, and hygiene gets better.

Genetic factors are considered to be less important. They determine the limit of individual potential, whereas environmental factors determine whether, and to what extent, this limit is attained. Assuming that this is true, one should expect secular trend extinction. This mainly concerns highly developed countries. It is estimated that populations in these countries have already attained their full genetic potential and therefore the improvement of environmental conditions will not result in crossing the limits determined by genetics. Investigations conducted by Bakwin and McLaughlin (1964) confirm this. On the basis of the research carried out on Harvard students living in rich families it was stated that no body height changes had occurred between the years 1930 and 1958. However, in the control group embracing subjects of limited means, an average increase of body height measurements of 40 mm was noticed.

Tanner (1962), a great authority in this field, having compared the results of research work conducted in the years 1960–1989, stated that all European countries, as well as the United States, Canada and Australia witnessed a gradual extinction of the trend during these years.

Similar conclusions were drawn by Roche (1979), who stated that investigations of juveniles in the United States carried out in the years 1962–1974 proved the absence of a secular trend. Pheasant (1996) assumes that the absence of a secular trend in the highly developed countries under investigation can be explained in two ways: first—all children have attained their full genetic potential, second—a certain percentage of children grow below their optimum environmental conditions, and therefore height measurements do not increase from one generation to another.

Németh and Eiben (1997) assume that phenotypic changes in the form of acceleration of growth and pubescence will no longer occur. In recent years some scientists (Fredriks et al. 2000) have rejected this assumption, and indicated that minor secular changes in highly developed countries continue to occur.

In Poland, and particularly in rural areas, secular trend can still be observed. It seems that theoretically one can still expect changes in Polish children in the next generation. Many scientists stress that the tendency of children's and juveniles' body measurements to increase does not occur with equal force in the whole population. For example, secular changes of height characteristics are more intensive in children growing up in the working class environment than in that of the intelligentsia (Bochenska 1979), and in children living in villages than in those living in towns and cities (Panek 1970; Krupinski et al. 1982; Boryslawski et al. 1988). The greatest inter-environmental differences in Poland are seen in the comparison of body height dimensions of children of the Warsaw intelligentsia with those of children living in villages whose parents have elementary education. Such large differences have never been observed between any other social groups. In children of the upper classes of society, the process of body height dimensions increasing undergoes extinction. Comparison of the body height arithmetic mean values of children living in the center of Warsaw, investigated in 1978 (Charzewski 1984; Nowak 1993), indicates that over the course of 14 years the stature dimension did not change in a significant way.

The values of the secular changes of body height at the age of seven to eight years conform to the dynamics of organism growth. Secular changes are differentiated according to particular age categories and they are more intensive at the age of accelerated development dynamics. The most significant differences are seen in the period of pubescence and they occur in boys between 12 and 14 years of age, and in girls between 11 and 12. As mentioned earlier, an important element of secular change is developmental acceleration, which is determined by the shift of the period of maximum growth, i.e. the pubertal spurt. When analyzing the yearly increases of the stature (B-v) and pubic height (B-sy) in children and juveniles aged 7 to 18, a process of pubertal spurt acceleration for both these characteristics can be clearly noticed (Tables 9.1 and 9.2).

TABLE 9.1   Annual Increase Values of Anthropometric Characteristics in Polish Boys from Big Town Populations in the Years 1966, 1978, 1992 and 1999 (in mm)

	STATURE (B-V)				PUBIC HEIGHT (B-SY)			
AGE	1966	1978	1992	1999	1966	1978	1992	1999
7–8	51	44	57	51	32	29	40	37
8–9	47	52	46	67	32	34	31	42
9–10	57	55	45	53	33	36	34	37
10–11	34	44	47	44	29	30	30	30
11–12	52	62	88	63	31	38	50	49
12–13	64	68	56	71	44	41	39	40
13–14	47	88	79	62	21	49	28	32
14–15	100	49	51	57	58	20	29	29
15–16	43	34	37	48	25	20	21	31
16–17	22	15	21	19	6	4	2	12
17–18	9	14	12	9	3	7	0	0

TABLE 9.2   Annual Increase Values of Anthropometric Characteristics in Polish Girls from Big Town Populations in the Years 1966, 1978, 1992 and 1999 (in mm)

	STATURE (B-V)				PUBIC HEIGHT (B-SY)			
AGE	1966	1978	1992	1999	1966	1978	1992	1999
7–8	49	54	59	60	43	35	45	46
8–9	56	45	56	58	25	31	34	30
9–10	57	54	48	51	38	35	25	35
10–11	55	73	75	71	37	44	50	42
11–12	57	70	63	55	27	42	34	27
12–13	66	45	44	61	38	24	17	35
13–14	29	38	45	29	14	16	18	15
14–15	27	14	18	14	7	8	16	5
15–16	6	17	12	12	8	6	7	16
16–17	5	10	5	6	1	7	2	9
17–18	3	14	5	3	1	10	4	10

In boys, the pubertal spurt of both body height and limb length shifted from the age of 14 to 15 years in 1966, to the age of 13–14 in 1978, and 11–12 in 1992. Since 1999 it has not been accelerating and its intensity has decreased, as confirmed by investigations by Palczewska and Niedzwiecka (2001). A similar trend is observed in girls—the pubertal spurt sped up from the age of 12–13 in 1966, to the age of 10–11 in 1992, and it does not continue to change at present. The shift of the pubertal spurt in those years was confirmed in the literature (Bielicki *et al.* 1981; Palus 1985).

Research by Hulanicka *et al.* (1990) indicates that the period of economic recession of the 1980s restrained the secular trend of children's and juveniles' body height increase. The author mentions:

> This phenomenon occurred most distinctly in children living in towns and cities. It concerned children born in the late 1970s. Their early childhood fell in the period of economic recession witnessed by the country. It is worth mentioning, however, that in this period children and the young from the best conditioned Warsaw environments showed further increase of body height. This phenomenon could confirm the specific conditioning of development in that large city.

Investigation of the menarche age carried out in Warsaw by Charzewski *et al.* (1984, 1998), which, similarly to body height, is one of the most precise indicators of the economic situation of a society, also proved the absence of the setback of the trend of body height dimensions growth, and, at the same time, a delay in the first menstrual period of girls in the early 1980s. Deceleration of the girls'maturation in the years 1978–1988 was also confirmed by investigations by Hulanicka *et al.* (1990) and in the years 1986–1997 by research by Charzewski *et al.* (1998). It should be stressed that deceleration of maturation was not observed at that time in any other European country. The interval between the investigations being the subject of this paper does not allow the separate determination of changes which occurred in each of the particular decades, i.e. in the 1980s and then in the period of political transformation of the 1990s. It can only be stated that the general trend of that period, covering the two decades observed by the researchers, is in conformity with the direction of the changes which have been noticed in Polish children and juveniles since the end of the previous century, i.e. acceleration of growth and bigger final body dimensions.

Slimming of the figure, manifested by the shift of the percentile ranges of the ratio of body mass to body height should also be perceived as an advantageous phenomenon. Children raised in better living and environmental conditions are slimmer and they have a thinner layer of subcutaneous fat tissue than children living in less favourable conditions (Kopczynska-Sikorska 1975; Niedzwiecka 1986).

Knowledge of the facts mentioned earlier is very important not only for anthropologists, but also for the specialists who design products for children and the young. The analysis conducted shows that in bygone years the Polish population was undergoing significant changes, which has resulted in the growth of body dimensions and different body proportions. Acceleration of the pubertal spurt has changed the pace of development, shifting the climax of the biggest growth to earlier years of life. Contemporary boys reach the climax at the age of 11–12 years, whereas girls reach it between 10 and 11 years of age. The pubertal spurt is a constant phenomenon and it occurs in all populations, though it differs in intensity and duration. On the basis of data available in the literature, it can be concluded that the rate of development of all European children is not the same. It is difficult to make an objective comparison since the data developed for particular countries were gathered in different years. For practical purposes, however, i.e. for the needs of ergonomics, the comparison of body height dimensions of boys and girls from selected European countries is presented in Tables 9.3 and 9.4.

Comparison of the mean values of the boys' and girls' body height (Tables 9.3 and 9.4) living in the following European countries: France (Norris and Wilson 1995), Germany (Flügel et al. 1986), Great Britain (Pheasant 1996), the Netherlands (Steenbekkers 1993), Hungary (Budavári and Eiben 1982), Norway (Waaler 1983), Poland (Nowak 2000), and Turkey (Dindar et al. 1989) indicates that the Dutch population achieves the highest stature dimensions in each age class. This applies to both boys and girls. Although the Dutch data only include children up to 12 years of age, the analysis of the rate of their physical development makes it possible to draw the conclusion that Dutch youngsters aged over 12 will also have the biggest values of stature among European juveniles. None of the populations compared in the tables, besides the Dutch one, achieves the height of 123 cm at the age of six and 157cm at the age of 12. Children from Poland, France, Britain, Norway, and Germany are tall. In general, body dimensions of boys and girls in these countries are similar. The biggest differences amount to about 2cm. The lowest body height is observed in girls and boys from Hungary and Turkey. These differences can result from both genetic predispositions and environmental conditions. It should be taken into account that some populations can still show the phenomenon of secular trend, all the more so as not all the data analyzed are up-to-date. The oldest data come from the 1980s and they concern the populations of Hungary (1982), Turkey (1989) and Germany (1986). The most recent data are of the Polish population and they have been developed as the prognosis for the year 2010 (Nowak 2000). Table 9.5 presents several somatic features developed for Polish boys and girls aged 6 and 18 years.

TABLE 9.3   Mean Values (x) of Stature (in cm) in Boys Aged 6–18 Years from Selected European Countries

AGE IN YEARS	EUROPEAN COUNTRIES							
	FRANCE	GERMANY	GR. BRITAIN	NETHERLANDS	HUNGARY	NORWAY	POLAND	TURKEY
6	118	120	117	122	118	119	121	116
7	127	128	123	128	122	124	125	120
8	130	131	128	134	127	130	132	125
9	136	137	133	141	133	135	137	130
10	140	144	139	146	139	140	142	135
11	145	145	143	151	143	145	146	140
12	153	152	149	156	148	150	155	145
13	158	163	155	–	154	156	161	149
14	164	168	163	–	162	164	167	157
15	169	171	169	–	–	170	174	165
16	173	176	173	–	–	174	178	–
17	174	–	175	–	–	177	180	–
18	176	–	176	–	–	–	181	–

TABLE 9.4 Mean Values (x̄) of Stature (in cm) in Girls Aged 6–18 Years from Selected European Countries

AGE IN YEARS	EUROPEAN COUNTRIES							
	FRANCE	GERMANY	GR. BRITAIN	NETHERLANDS	HUNGARY	NORWAY	POLAND	TURKEY
7	125	128	122	129	120	123	125	119
8	134	131	128	134	126	128	131	124
9	137	137	133	139	133	134	137	129
10	143	144	139	147	138	139	142	134
11	148	148	144	151	144	145	149	140
12	161	153	150	157	151	151	156	145
13	162	160	155	–	156	157	160	151
14	162	161	159	–	157	162	165	157
15	162	163	161	–	–	164	166	164
16	162	165	162	–	–	165	166	–
17	163	–	162	–	–	166	168	–
18	163	–	162	–	–	–	169	–

TABLE 9.5   Anthropometric Measured Data (in mm) of Polish Boys and Girls Aged 6 and 18 (Nowak 2000)

DIMENSION	BOYS				GIRLS			
	6 YEARS		18 YEARS		6 YEARS		18 YEARS	
			PERCENTILES					
	5	95	5	95	5	95	5	95
Stature	1139	1285	1708	1921	1132	1285	1592	1811
Eye height	1021	1158	1582	1775	1015	1163	1492	1630
Acromion height	897	1029	1438	1591	885	1015	1329	1465
Suprasternal height	903	1029	1392	1584	886	1026	1290	1482
Elbow height	700	818	1108	1242	688	802	1030	1152
Pubic height	572	665	892	1025	574	663	855	956
Head and neck height	232	252	312	332	238	255	297	324
Trunk height	335	368	498	564	320	367	450	530
Thigh length	281	301	440	468	273	305	412	449
Knee height	288	361	456	556	298	355	445	503
Upper extremities length	434	574	717	876	427	569	650	805
Arm length	173	233	301	372	175	233	275	338
Forearm length	144	199	233	292	137	192	207	268
Hand length	114	138	177	206	112	140	163	193
Arm overhead reach[a]	1309	1531	2113	2345	1271	1491	1900	2158
Arm reach down[a]	428	525	701	831	419	518	653	777
Arms span[a]	933	1164	1473	1774	949	1172	1355	1605

[a] Measurement taken with the hand clenched.

9.3 APPLICATION OF ANTHROPOMETRY TO THE DESIGN OF SCHOOL FURNITURE

School furniture is a special case. It belongs to that group of products which can considerably influence the development of a young, rapidly changing organism. Pupils today spend more and more time in the sitting position, both at school and at home, using a computer. More and more researchers (Balaqué *et al.* 1993; Jarosz 1993; Nowak 1993; Storr-Paulsen *et al.* 1994; Marschall *et al.* 1995; Mandal 1997; Paluch 1999) unfortunately confirm the fact that most school furniture does not correspond with the children's dimensions, and its faulty construction forces harmful body posture. Sitting for a long time in a "bad" position can cause defects and diseases of the vertebral column. The most-often observed errors include seats and tables which are too low or too high. When the height of the seat is not in conformity with the popliteal height measurement and exceeds its value, pressure is exerted by the front edge of the seat on the blood vessels of the thighs (the seat buries into the thighs). This causes circulatory disturbances and can lead to numbness and swelling of the lower limbs. Too low a seat also has its faults. Most often children, especially smaller children, put their shanks under the chair.

In the case of youths, when the disproportion between the seat height and the extremity length is much more significant, the knees go up. The thighs and the trunk form an acute angle and internal organs are squeezed. Raised thighs do not rest on the seat and the trunk weight is mainly carried by the ischiadic tubers.

In young people whose adipous layer is thin, this is very painful. Another, also disadvantageous, situation occurs when the seat is too deep. In this case, to avoid the seat edge burying into the area of the popliteal fossa, the child subconsciously moves forward. In consequence, the vertebral column lacks lumbar support, which is exceptionally harmful for its proper functioning. Dimensions of the tables at which children work at school and at home should also comply with anthropometric measurements coherent with an adequately selected chair. Too low a table forces both back separation from the backrest of the chair and inclination of the trunk, which assumes a C-shape. As a result the trunk loses its physiological curve. Too high a table forces unnatural arm-heightening and causes too big static muscle tonus in the acromial girdle.

Anthropometric measurements used for determining school furniture parameters are given below.

CHARACTERISTICS	FURNITURE ELEMENTS
Popliteal height[a]	Seat height
Popliteal depth	Seat depth
Buttocks breadth	Seat breadth
Lumbar lordosis height[a]	Bottom edge of the backrest
Infrascapular height[a]	Top edge of the backrest
Elbow height[a]	Table top height
Elbow span	Table top length
Thigh thickness	Bottom edge of the table height (place for legs under the table)

[a] Characteristics measured from the seat level.

About 20 mm (for shoes) should be added to the popliteal height measurement and about 10 mm (for clothes) to the remaining measurements. While designing the top and the bottom height of the table surface, 10 to 25 mm should be added so that the elbow is placed slightly above the work plane and the thighs have enough space. The characteristics listed above are used in the development of standards. In recent years, in many countries including Sweden, Denmark, Germany, France, and Poland, scientists and teachers have observed that obligatory standards for school furniture do not comply with current dimensions of the young. The Association of Swedish Teachers indicated that body height dimensions in the ISO Standard (1990) should be increased by 200 mm. The Association of Danish Schools suggested the same. On the basis of the studies carried out by the Institute of Industrial Design (Nowak 2000) it was stated that the Polish Standard PN-ISO 5970 did not ensure appropriate conditions for the tallest young people. Thus the Institute of Industrial Design suggests adding one more condition, the highest dimension, to the valid ISO standards. This suggestion complies with the European Prestandard DD ENV 1729.

Another important aspect of school furniture standardization is the method of arranging furniture dimensions. The ISO Standard assumes body height dimension as the criterion of division. According to the studies conducted on Polish children, popliteal height seems to be a better criterion. The dimension of this feature consists of several characteristics including the lengths of trunk, thigh, and shank. These characteristics show slightly different rates of growth. For example, in Polish boys from the age of 11 and in Polish girls from the age of nine, knee height is more advanced in its development than body height, and in the period from 8 to 18 years of age, knee height increases faster than thigh length. Moreover, a child before the period of puberty can have the same body height as his/her peer, but smaller thigh or shank measurements. Thus, a chair adjusted to the child's dimensions according to the body height criterion may be too low. Therefore the method of furniture designing in compliance with the body height characteristic seems to be controversial. This problem was first noticed by Austrians, who assumed popliteal height as the primary determinant of furniture dimensions (Standard ONORM1650–1987). The European Prestandard mentioned earlier takes into account this feature, as well as the stature characteristic, as the basis for standardization.

Another controversial question is determination of the angle between the seat and the backrest. Recently there have been more publications which favour the sitting position with thighs sloped downward, where the angle between the trunk and the thigh is more than 90°. This position is secured by the use of forward sloping seats. Those who support the idea of this kind of a seat indicate that it minimizes vertebral column loading in its lumbar segment and facilitates free blood circulation due to reduced pressure exerted on internal organs and thighs.

Investigations conducted at the Institute of Industrial Design (IWP) in Warsaw on a group of 64 children aged 8–19 indicated that, despite so many advantages, a seat sloping forward by the angle of 4° is not accepted by all children—36% of the subjects evaluated it as uncomfortable (Nowak *et al.* 2002). The children indicated that they had the feeling of slipping down (in spite of the rough surface) and in order to prevent this they were "digging in their heels." Possibly the children were used to the traditional position and thus the position of the so-called "obtuse angle" did not suit them. The European Prestandard mentioned above recommends setting the angle within the limits of 3° to 3°. The Prestandard was developed with consideration for the previously-mentioned suggestions put forward by research teams, i.e. it includes enlarged dimensions of school furniture, popliteal height as an indicator of furniture adjustment and the forward sloping seat.

9.4 TWO-DIMENSIONAL MANIKINS OF CHILDREN: MODELS FOR DESIGN

On the basis of anthropometric data, anthropometric models that imitate the silhouettes of children aged 6 to 18 years have been developed at the Institute of Industrial Design as a result of close cooperation between an anthropologist and a designer.

The following criteria were taken into consideration while preparing the manikins: conformability of functional measurements with current anthropometric data, simplicity of the use of models, of their removal and storing and the lowest costs of production. In order to meet these criteria it was necessary to reconcile many factors important both for anthropometry and design. It is obvious that even the best model is a static arrangement—it cannot fully present the dynamics of the human body, and especially the dynamics of the changing ontogeny of a child's body. To achieve data synthesis the idea of differentiating manikins in respect of sex was abandoned. The analysis of the development of body height as well as of other height features of boys and girls indicates that the development of these features in boys and girls up to ten years of age is similar and that differences of somatic features are statistically insignificant (up to 8 mm) and can be neglected for design purposes (Nowak 1989).

The manikins were not prepared for each age class but were divided into groups consisting of several age classes. It was very difficult to separate the groups since it was impossible to take into account the dynamics of development with the division resulting from the necessity of designing furnishings separately for children in kindergartens, primary schools and secondary schools simultaneously. The period between 11 and 15 years of age was the most controversial one. This is a period of rapid changes in a child's body due to puberty. This period differs slightly between boys and girls. The pubescent spurt is a sign of puberty and results from the hormonal, functional and morphological changes that are taking place in a child's body in preparation for puberty. Although the pubescent spurt is common to both boys and girls, its effects differ in intensity and persistence in individual children.

In the case of boys, the spurt occurs on average at the age of 14; it is about two years earlier, and less intensive, in girls. The differences in dimensions between adult males and females result to a certain extent from the differences of their pubescent spurt. Since the girls' pubescent spurt starts earlier, it is less intensive and as a result the process of body development is completed earlier. On average, girls reach puberty at the age of 16 and, at the same time, their bodies cease growing; in the case of boys, their development process can continue up to 21 years of age.

In view of this, the groups of boy and girl manikins were divided differently, although the divisions were made according to successive stages of child development. These covered post-infantile age (one to three years), pre-school age (four to six years), younger school age (seven to ten years), puberty (11–14 years) and juveniles (15–18). This division corresponds with the ages of children attending nursery, kindergarten, primary and secondary schools. Thus, the following seven groups were selected according to age: I: one to three-year-old boys and girls; II: four six-year-old boys and girls; III: seven to ten-year-old boys and girls; IV: 11–14-year-old boys; V: 15–18-year-old boys; VI: 11–13-year-old girls; VII: 14–16-year-old girls.

The set of manikins was prepared using the values of the 5th and 95th percentiles, to allow the application of threshold percentiles in design. The set consists of 26 models presented in two views: a side view, where figures are in the sagittal plane, and a top view, where figures are in the transverse plane. Each model gives information on sex, age and its percentile value. The manikins are made of a stiff material, plexiglass, and in a scale of 1:5, and have holes which enable the placing of the figure in a given position or the movement of a given segment of the body. The manikins are fixed by sticking a compass leg or a pin into a hole on the top of the head or on a foot. Movement capacities are possible thanks to holes representing the axis of rotation of the following joints: shoulder, elbow, wrists, hip, knee, ankles, as well as head motions (upper and lower joints of head), neck motions (junction of neck and thoracic segments of spine) and trunk motions (junction of thoracolumbar segment). Flexion and extension movements in all these joints can be obtained by means of the manikin developed in the sagittal plane, while the manikin developed in the transverse plane allows the determination of the abduction and adduction movements of limbs and head movements. The model in the transverse plane is a simplified shape of a child's figure seen from above. The left side consists of the shoulder, upper limb and breast; the right side consists of the thigh, hip, and abdomen. By rotating the manikin along the axis of symmetry, one can obtain the contour of the whole body. Contours of upper limb and reach dimension are obtained by putting the model into the sagittal plane.

Figure 9.1 shows the application of manikins in the ergonomic analysis of furniture. If the analysis is undertaken while still in the design stage, this

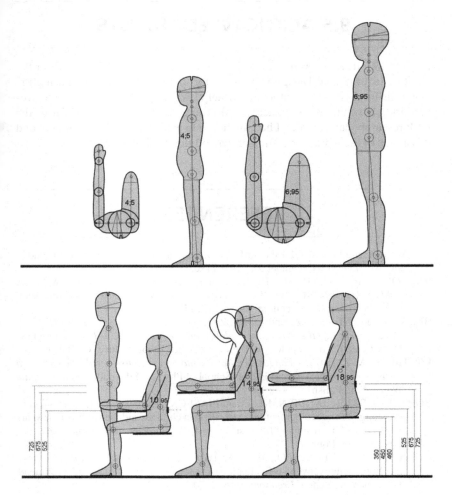

FIGURE 9.1   Application of manikins in ergonomic analysis of furniture.

allows any problems in the functional dimensions of furniture to be revealed and corrected.

Besides the manikins and the publication, i.e. the Anthropometric Atlas of the Polish Population—Data for Design (Nowak 2000), Polish designers have a third tool to assist in designing the environment for children and youths: it is the set of anthropometric data based on the prognostic data (Nowak 2000), also developed at the Institute of Industrial Design in a digital form and distributed to designers on a CD.

9.5 ACKNOWLEDGMENTS

I would like to express my gratitude to my colleagues and friends from the Institute of Industrial Design in Warsaw for their assistance in preparing the materials for this work—without their help I would not have been able to prepare the above chapter so quickly and efficiently. I am also thankful to the IWP research teams, which I had the honor to lead, for their commitment and assistance in the work, the results of which I present in this book.

REFERENCES

BAKWIN, H. and McLAUGHLIN, S.M., 1964, Secular increase in height. Is the end in sight? *Lancet*, 13(December 5), 1195–1196.
BALAQUÉ, F., DAMIDOT, P., NORDIN, M., PARNIANPOUR, M. and WALDBURGER, M., 1993, Cross-sectional study of the isometric muscle trunk strength among school children. *Spine*, 18, 1199–1205.
BIELICKI, T., WELON, Z. and WALISZKO, A., 1981, *Changes in the Physical Development of the Young in Poland in the Years 1955–1978* (Wroclaw: Department of Anthropology of the Polish Academy of Science) (in Polish).
BOCHENSKA, Z., 1979, *Changes in Man's Ontogeny Against Secular Trends and Social Differences* (Cracov: Higher School of Physical Education), *Monographic Works*, 5 (in Polish).
BORYSLAWSKI, K., KRUPINSKI, T. and PIASECKI, E., 1988, Environmental conditioning of the development of lower Silesian children and the young in the years 1986–1990. The results of the first study. *Anthropological Works and Materials*, 109, 7–14 (in Polish).
BUDAVÁRI, E. and EIBEN, O., 1982, Az iskolai bútorok értékelése a tanulók testméreteinek függvényében (Evaluation of school furniture with regard to Students Body measurements). *Ergonomia*, 15, 2, 70–77.

CHARZEWSKI, J., 1984, *Social Conditioning of Physical Development of Children Living in Warsaw* (Warsaw: Academy of Physical Education) (in Polish).

CHARZEWSKI, J., LEWANDOWSKA, J., PIECHACZEK, H., SYTA, A. and LUKASZEWSKA, L., 1998, Menarche age in Warsaw girls 1986–1997. *Physical Education and Sport*, 1, 61–66 (in Polish).

DINDAR, H., YUCESAN, S., OLCAY, I., OKUR, H., KILICASLAN, S., ERGOREN, Y., TUYSUZ, C. and KOCA, M., 1989, Physical growth measurements of primary school children living in Adana, Turkey. *Turkish Journal of Pediatrics*, 31, 45–56.

FLÜGEL, B., GREIL, H. and SOMMER, K., 1986, *Anthropologischer Atlas* (Berlin: Verlag Tribüne).

FREDRIKS, A.M., VAN BUUREN, S., BURGMEIJER, R.J.F., MEULMEESTER, J.F., BEUKER, R.J., BRUGMAN, E., ROEDE, M.J., VERLOOVE-VANHORICK, S.P. and WIT, J.M., 2000, Continuing positive secular growth change in the Netherlands 1955–1997. *Pediatric Research*, 47(3), 316–323.

HULANICKA, B., BRAJCZEWSKI, C., JEDLINSKA, W., SAWINSKA, T. and WALISZKO, A., 1990, *Large Town — Small Town — Village, Differences in the Physical Development of Children in Poland* (Wroclaw: Department of Anthropology of PAN) (in Polish).

JAROSZ, E., 1993, Final anthropometric standards for the needs of working furniture standardization. *Wiadomosci IWP*, 3, 69–70 (in Polish).

KOPCZYNSKA-SIKORSKA, J., 1975, Main tendencies of the somatic development of children and the young in Poland in the last two decades. *Issues of Medicine of Development Age*, 6, 79–88 (in Polish).

KRUPINSKI, T., BELNIAK, T., BORYSLAWSKI, K., KLONOWSKA, E. and KORNAFEL, D., 1982, Variability of somatometric characteristics of children and the young living in villages in the selected regions of the country on the basis of the measuring of two local populations repeated after 10 years. *Anthropological Works and Materials*, 102, 101–114 (in Polish).

MANDAL, A.C., 1997, Changing standards for school furniture. *Ergonomics in Design*, 5(2): 28–31.

MARSCHALL, M., HARRINGTON, A.C. and STEELE, J.R., 1995, Effect of work station design on sitting posture in young children. *Ergonomics*, 38(9), 1932–1940.

NÉMETH, Á. and EIBEN, O.G., 1997, Secular growth changes in Budapest in the 20th century. *Acta Medica Auxologica*, 29, 5–12.

NIEDZWIECKA, Z., 1986, Physical development of Warsaw children living in good environmental conditions. *Physical Education and Sport*, 30(3), 39–57 (in Polish).

NORRIS, B. and WILSON, J.R., 1995, *Childata. The Handbook of Child Measurements and Capabilities — Data for Design Safety* (Nottingham: Institute for Occupational Ergonomics, University of Nottingham).

NOWAK, E., 1989, Two-dimensional manikins of children: Models for design. *Applied Ergonomics*, 20.2, 136–139.

NOWAK, E., 1993, *Anthropometric Data for Designing a Pupil's Workstand. Prace i Materialy IWP*, 148 (Warsaw: Institute of Industrial Design) (in Polish).

NOWAK, E., 2000, *Anthropometric Atlas of the Polish Population — Data for Design* (Warsaw: Institute of Industrial Design).

NOWAK, E., SYCZEWSKA, M., KALINOWSKA, M., KALKA, E., JAROSZ, E., LAPACZEWSKA, K., WRONSKI, S., OPORSKI, J., KSIEZYC, D. and LEPIANKO, J., 2002, *Environment Shaping — School Furniture* (Warsaw: Institute of Industrial Design), typescript, unpublished paper (in Polish).

PALCZEWSKA, I. and NIEDZWIECKA, Z., 2001, Indicators of the somatic development of Warsaw children and the young. *Medicine of Development Age*, V(Suppl. 1–2): 18–118. (in Polish).

PALUCH, R., 1999, Ergonomics at school. Application of ergonomics. *Centrum Zastosowania Ergonomii*, 2, 45–59.

PALUS, D., 1985, Differentiation of children's and the youth's body proportions as the basis for school furniture standardization. *Anthropological Works and Materials*, 105, 15–58 (in Polish).

PANEK, S., 1970, Secular trend in the growth of Polish urban and rural children, examined in 1956 and 1966. *Anthropological Works and Materials*, 79: 3–29 (in Polish).

PHEASANT, S.T., 1996, *Bodyspace: Anthropometry, Ergonomics and Design* (London: Taylor & Francis).

ROCHE, A.F., 1979, Secular trends in stature, weight and maturation. *Monographs of the Society for Research in Child Development*, 179, 44(3–4), 3–27.

STEENBEKKERS, L.P.A., 1993, *Child Development Design Implications and Accident Prevention* (Delft: Faculty of Industrial Design Engineering, Delft University of Technology).

STORR-PAULSEN, A. and AAGAARD-HENSEN, J., 1994, The working positions of schoolchildren. *Applied Ergonomics*, 25(1), 63–66.

TANNER, J.M., 1962, *Growth at Adolescence, with a General Consideration of the Effects of Hereditary and Environmental Factors Upon Growth and Maturation from Birth to Maturity* (Oxford: Blackwell Scientific Publications).

WAALER, P.E., 1983, *Anthropometric studies* in *Norwegian children. Acta Paediatrica Scand Suppl. 1983, 308*: 1–41.

FURTHER READING

Bravo, G., Braganca, S., Arezes, P. M., Molenbroek, J. F. M., & Castellucci, H. I. (2018). A literature review of anthropometric studies of school students for ergonomics purposes: Are accuracy, precision and reliability being considered? *Work*, 60(1), 3–17.

Cheng, I. F., Kuo, L. C., Lin, C. J., Chieh, H. F., & Su, F. C. (2019). Anthropometric database of the preschool children from 2 to 6 Years in Taiwan. *Journal of Medical and Biological Engineering*, 39, 552–568.

Fryar, C. D., Gu, Q., Ogden, C. L., & Flegal, K. M. (2016). Anthropometric reference data for children and adults; United States, 2011–2014.

Inokuchi, M., Matsuo, N., Takayama, J. I., & Hasegawa, T. (2019). National anthropometric reference values and growth curves for Japanese children: History and critical review. *Annals of Human Biology*, *46*(4), 287–292.

Kennedy, S., Smith, B., Sobhiyeh, S., Dechenaud, M. E., Wong, M., Kelly, N., Shepherd, J. & Heymsfield, S. B. (2022). Digital anthropometric evaluation of young children: Comparison to results acquired with conventional anthropometry. *European Journal of Clinical Nutrition*, *76*(2), 251–260.

Anthropometry for the Needs of the Elderly

10

E. Nowak

10.1 INTRODUCTION

Many theories have been developed in the search of the aetiology of the ageing process in living creatures; however, none of them adequately explains this phenomenon. Only fragmentary mechanisms of this process course are determined, for example, Harman's theory concerning free radicals. Generality and, up to the present, irreversibility of ageing indicate that it is conditioned genetically (particular number of fibroblasts division determined for every man), but it also depends on environmental conditions.

Numerous attempts to divide man's life into phases, including the ageing period, give rise to much controversy. Conventional character of the division results in the fact that the boundaries of the defined phases are not clear. According to the criteria used by biologists and physicians, it can be assumed that at present the ageing period falls on 60–90 years of age, and two sub-periods are distinguished: early old age—the so-called second age—between 60 and 70 years of age, and mature old age from 75 to 80. Venerable old age is stated after the 90th year of age up to the end of one's biological life.

An organism ageing is a complex process. Rapid regressive (involutional) morphological and functional changes take place in the cells and organs in ontogenesis; they result in the reduction of fitness and psychophysical

DOI: 10.1201/9781003459767-10

efficiency of the organism. No other period of life is characterized by such significant interindividual differences as the old age. Elderly people are not a homogenous group. Members of this collectivity have different features of character, different likes and dislikes, health state, family situation and living conditions, they live different lives. At present, due to medicine progress and better sanitary conditions, the majority of countries witness the shift of old age. Nevertheless, the ageing of human populations has become a fact. The World Health Organisation (WHO) estimates that in the year 2025 the average life span will amount to 80 years in 26 countries. Simulations developed for the UK forecast that in 2021 people over middle age will prevail in the population and a remarkable growth in the number of people aged more than 80 will be observed; at the same time the young adult population will be reduced (Coleman 1993). Similar demographic prognoses have been formulated for Sweden, where elderly people aged over 50 will be in the majority, and the number of younger age groups will decrease.

Unfavourable demographic trends, characteristic of highly-developed European countries, also concern Poland, where constant growth in the number of no longer economically active people and their increasing percentage in the total number of inhabitants are also observed. The ageing process of the Polish population takes a gradual course. Demographic prognoses developed for Poland for the years 1995–2020 assume that the group of elderly people which amounted to 5 300 000 people in 1995 will include as many as 6 100 000 people by 2010. The years 2011–2020 will witness an increase of another 1 800 000. Thus, in Poland, the number of people of pensionable age forecast for the year 2020 will be approximately 7 900 000 (Report of the Ministry of Economy 1997). According to long-term prognoses, in the year 2045 over 10 500 000 people will come up to pensionable age (Demographic Year-Book, Central Bureau for Statistics 2000). Through the research on the Polish population, it has been shown that elderly women considerably prevail over elderly men in respect of numbers. This phenomenon results from the higher mortality level of men (especially those younger) and their shorter average life span.

Changes in mean life span are one of the factors which significantly influence the process of the population's ageing. According to data from the Central Bureau for Statistics the average life length of Poles is constantly increasing. In 1996 the average life expectancy in Poland was 68.1 years for men and 76.6 years for women. The growing number of elderly people presents a challenge to secure appropriate living conditions for this group of inhabitants.

Anthropometry as an important part of ergonomics also contributes to this. The task of anthropologists is to describe the somatic characteristics of an elderly person and changes that occur in the structure and body proportions of the ageing organism. By adjusting the articles of daily use, appliances, and interior furnishings to the dimensions and physical predispositions of the elderly, ergonomics not only provides this group of people with the facilities for living independently, but also contributes to the increase of their life comfort and often prevents them from dangerous accidents. UK yearly reports show that 15% of all home accidents recorded in England in 1995 concerned people aged over 65.

Investigations undertaken by ergonomists in the UK, the Netherlands, Germany, Japan, the USA, Canada, and Scandinavia indicate that the home environment exposes the elderly to numerous accidents that could lead to permanent injury. According to the report of the Dutch Institute for Consumer Security 74 000 household accidents were recorded in the Netherlands in 1980; 24 000 of them ended in death and 50 000 in injuries. These proportions were particularly disadvantageous for elderly people: 75% of deaths and 25% of the loss of efficiency in the whole population (Molenbroek 1987). At the same time, the author indicates that women cause twice as many accidents as men. This high accident rate results in enormous costs of medical treatment and financial recompensations; thus the adjustment of work and life environments of the elderly to their anthropometric measurements which change with age, and their decreasing physical efficiency, is so important.

10.2 ANTHROPOMETRIC RESEARCH OF THE ELDERLY

10.2.1 Somatic Characteristics

One of the characteristics which change significantly with age is stature. Source materials cited in the world literature indicate that this characteristic starts to decrease after the age of 25–35. Generally, it is assumed that body height decreases gradually by 1–2 cm per decade, and after passing 90 years of age, the process of body height shortening becomes more rapid.

Body height decrease results, among others, from numerous changes that occur in osteoarticular and muscular systems. As a consequence of the ageing process, flattening of intervertebral bodies and cartilages, as well as deepening of chest kyphosis and reduction of lumbar lordosis, occur.

These are accompanied with the decline of muscular tension in the chest and abdomen area, downward shift of ribs, and shortening of the trunk which takes the shape of the letter C. Lower extremities become shorter as well. As a result of deeper excavation of the femur's head in the hip acetabulum, the angle between the neck and the body of the femoral bone lessens. These changes are usually seen in people aged over 65. The structure of the chest also changes. Its mobility weakens, its depth increases, and its shape—flat in a young man—becomes bulgy with age.

The morphological alternations described above, which decide on the appearance of an elderly person, result in a different body posture including the following main characteristics: the trunk inclined forward, intensified chest kyphosis (often leading to "the hump back of the old man"), lower extremities bent compensatorily in the hip and knee joints. As a result, elderly people are usually short, have short lower extremities, and small movement ranges (Nowak 2000). The differences in the measurements of the above characteristics in comparison with those of adults can be significant. The review of the data of the European population (Fidanza et al. 1984; Borges 1989; Delarue et al. 1994; Jarosz 1998; Smith et al. 2000; Nowak and Kalka 2003) indicates that these differences can amount from approximately 40 to 100 mm. The data on the stature of the populations of men over 65 from different countries are compared in Table 10.1. Though the compared groups are not homogeneous in respect of the men's social and economic status and health condition, the size of the groups, and the time of the investigation, they constitute a certain source of information for a designer regarding anthropometric characteristics of a given group.

Body mass does not change so explicitly. Generally, in developed countries, the growth of body mass occurs in both women and men of middle-age. In advanced age, after passing the age of 65 years, body mass also changes, but these changes are slightly different in men and women. As the source data indicate (Borges 1989; Delarue et al. 1994; Ravaglia et al. 1997; Smith et al. 2000; Nowak and Kalka 2003), elderly European women over 65 years of age do not get slimmer, but even put on weight and are heavier than adult women by 1–10 kg. In the men from the compared populations (Table 10.2)

TABLE 10.1 Comparison of the Mean Values of Stature of Males and Females Aged over 65 and Adults from Selected European Countries (in mm)

| COUNTRY | MALES | | | FEMALES | | |
	ELDERLY	ADULT	AGE	ELDERLY	ADULT	AGE
Finland	1695		75	1558		75
France	1683	1748	65–69	1580	1652	65–69
Great Britain	1711	1755	65–74	1583	1620	65–74
Netherlands	1737	1795	65–69	1611	1650	65–69
Italy	1625	1728	65–70	1507	1610	65–70
Poland	1698	1778	65–75	1570	1634	60–80
Sweden	1760	1740	70	1620	1640	70

TABLE 10.2   Comparison of the Body Weight of Males and Females Aged over 65 and Adults from Selected European Countries (in kg)

COUNTRY	MALES			FEMALES		
	ELDERLY	ADULT	AGE	ELDERLY	ADULT	AGE
Finland	74	73	75	68	58	75
France	77	73	65–69	59	58	65–69
Great Britain	78	80	65–74	68	67	65–74
Netherlands	78	76	65–69	69	65	65–69
Italy	71	74	65–70	60	57	65–70
Poland	79	76	65–75	69	66	60–80
Sweden	77		70	64	59	70

only elderly Britons and Italians are lighter than younger subjects. In general, it is assumed that the significant loss of body mass occurs after 80 years of age.

As the ageing of the organism proceeds, the whole motor units, together with neurons, disappear. Up to 90 years of age, muscle mass lessens by approximately 30% in relation to the maximum mass of the same person in his or her adulthood. Power of the muscles weakens and disturbances in the static and dynamic muscle activity occur more often; this leads to the muscles overloading.

The skeleton and joints structure also changes with age. Demineralization of bones and the loss of their elastic elements contribute to the limitation of their mechanical resistance which leads to injuries and pathological states of the motor organ. The skeleton becomes increasingly less resistant to external forces. As a result of the alternations occurring in the organism, changes in body shape and proportions take place.

10.2.2 Elements of Anthropometry in Workspace Design

Sociological studies indicate that elderly people perceive self-independence as the main indicator of their state of being satisfied with life. This means first of all living in an own flat or house with the possibility of running the house on their own. It is also important to stay in the same home, with familiar furniture to which the owners often become very attached. These factors are of great importance for a designer whose task is to arrange such interiors. If elderly people have to move to another place, it should be designed in conformity with the principles of ergonomics, and, at the same time, with respect for individual preferences. One of the main rules is to adjust spatial structures and furnishing elements to the physical structure of man.

As indicated in Table 10.1, elderly people have smaller stature measurements than those middle-aged. They are also characterized by, among others, smaller length and height measurements of the particular body parts and considerably lower values of functional measurements, including reach, which is very important for the design of spatial structures. The fact that the reach values decrease with age was proved, among others, by the research which embraced the populations of Poland (Jarosz 1998), the Netherlands (Molenbroek 1987) and the UK (Pheasant 1996). The reach measurements are important while designing interior furnishings, mainly when arranging shelves, switches, and handles.

Taking into consideration anthropometric measurements of the Polish population, biomechanical aspects and the type and frequency of manual activities performed, three working reach zones can be distinguished (Nowak and Kalka 2003). The measurements of these zones are presented in Figure 10.1.

The space between the acromion height and elbow height was assumed to be optimum. All kinds of switches, push buttons, and handles, as well as other manually operated elements, should be located in this zone along with products which are often used. For elderly people in the standing position this zone, measured from the floor level, extends from 1200 to 850 mm. The less frequently used elements mentioned above can be placed in the lower zone, between 850 and 610 mm. Fixing them below this level exposes an elderly person to back bone and knee joint overloading as a result of bending. Reduced joint elasticity, as well as degenerative diseases of the arthroligamentous system, causes additional problems. According to Kirvesoja *et al.* (2000), the level of shelves in the living interiors meant for elderly people below 300 mm is inadmissible. The above study indicates that it is easier for elderly people to reach for a product placed on a shelf located slightly above the shoulder joint level than to bend for it. The authors recommend that the upper limit of the zone for storing products should not exceed 1600 mm. Investigations carried out on the Polish population revealed the same.

With the use of the same principles that define work comfort, reach zones for elderly users working in the sitting position (Figure 10.2) were determined.

Referring to the recommendations of some authors (Pheasant 1996; Kirvesoja *et al.* 2000; Kothiyal 2001), the dimension of the seat height equal to the popliteal height of women of the 5th percentile was defined. The upper reach in this position is equal to 1280 mm and the lower reach to 290 mm. The optimum zone extends from 810 mm to 470 mm (a shoe allowance of 20 mm should be added to all height measurements). Additionally, reach zones for the elderly using a wheelchair were calculated and compared with those of the disabled. The standard height of a wheelchair—equal to 530 mm—was accepted. For both groups under investigation, the ability to reach down is identical. The

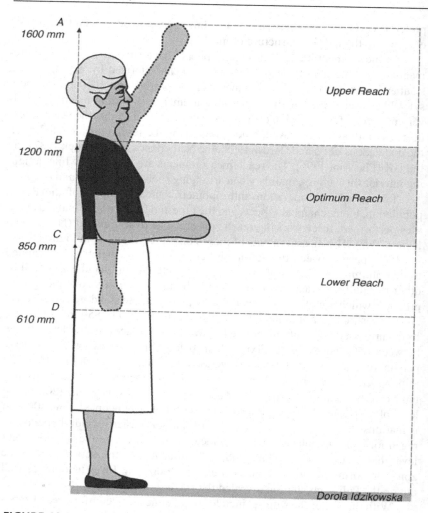

FIGURE 10.1 Reach zones of the elderly in the standing position.

upper limits of the bottom and optimum zones differ by 10 mm. The most significant difference concerns the upper reach and amounts to 58 mm to the disabled's disadvantage. Table 10.3 compares the reach zones of the elderly and the disabled on the assumption that both groups are wheelchair users.

Shelves for products storing, designed for the interiors meant for the elderly and the disabled, as well as other elements which require frequent manipulation, should not be placed beyond the zone of 1470 mm under any circumstances.

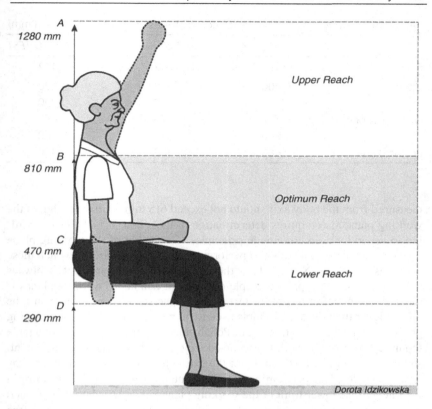

FIGURE 10.2   Reach zones of the elderly in the sitting position.

In the case of the disabled, the level should not exceed 1412 mm. For both groups, the lowest reach down amounts to 480 mm. A disabled person is not able to reach anything placed below this level. He or she cannot lean sideways, firstly due to restricted body mobility, and also because movements of the extremities are impeded by the side-rests. A wheelchair user can neither pull him/herself up, nor bend forward. Any attempt to change body position causes the shift of the centre of gravity, leading to a loss of stability which can result in an accident. Therefore, it is also important to determine the reach zone in the horizontal plane. Dimensions of this zone are essential for the design of interiors and working planes which require reaching sideways and forwards. Forward reach zone measured from the backrest, determined by Nowak (2003) for the Polish population, amounts to approximately 620 mm. This constitutes the limit for placing all kinds of objects, devices and switches, as well as shelves, drawers etc. in the frontal plane. From the side, the dimension

TABLE 10.3   Comparison of the Reach Zones of the Elderly and Disabled (in mm)

	THE ELDERLY	THE DISABLED	COMBINED TEST
	1470	1412	1412
Upper zone	1000	990	990
	1000	990	990
Optimum zone	660	670	660
	660	670	660
Bottom zone	480	480	480

measured from the body axis should not exceed 615 mm. Correct design of the working plane also requires determination of its height. It is defined according to generally used ergonomic principles. The height of the working plane depends on the manual activities performed and on the position in which these activities are done. Nevertheless, the elbow height measurement is always the determinant. A typical example of different heights of working planes is the kitchen. A buffet meant for fast eating in the standing position would be much higher than the top of a table used to prepare meals, cutting or chopping products, or peeling vegetables. For Polish elderly women, the height of a table top meant for kitchen work (with consideration to elbow height measurement) should amount to 700–900 mm. For the sitting position, which is preferred by elderly people while performing various manual activities, including work in the kitchen, the lower limit of the working plane height (A) should be located above the maximum measurement of thigh thickness; an extra 25 mm of free space should be added. The upper limit of the working plane height (B) should be approximately 25 mm above the elbow height (also measured from the seat plane). In the case of Polish elderly women, the A dimension should not be less than 175 mm and the B dimension should not exceed 280 mm (Figure 10.3).

In order to secure functionality of home interiors, it is also very important, especially for elderly people, to determine not only appropriate measurements of working planes and zones meant for storing products, but also to arrange furniture and equipment in a proper way. One of the specific working places in which numerous ergonomic problems have to be solved is the kitchen. According to the principles of ergonomics, the main working planes for the right-handed should be placed in the following order (clockwise): sink–basic working plane–cooker–another working plane for placing products. Kitchen spaces can vary in their shapes and sizes. Although a larger kitchen can be equipped with more working planes and more kitchen appliances, it can be difficult for an elderly person to prepare meals in such a kitchen since this requires covering much larger distances. Figure 10.4 presents three different shapes of kitchen in which the equipment is arranged in line or in the shape

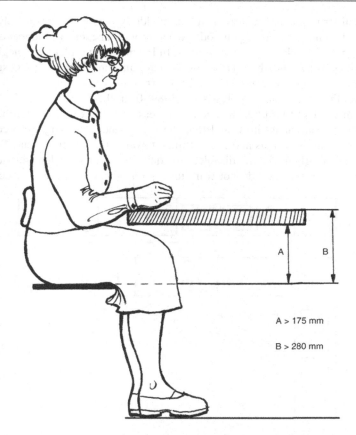

FIGURE 10.3 Working plane height appropriate for elderly women in the sitting position. A—lower limit of the table top (space for legs); B—upper limit of the table top.

of the letter L or U. The U option is the best one for elderly people since three basic appliances: refrigerator, sink, and cooker unit constitute the so-called "working triangle" that requires covering the shortest distance.

10.2.3 Elements of Anthropometry in Clothing Design

For the elderly, clothing is one of the important, though still underestimated, products which can significantly influence the quality of life for this group of people. First of all clothing should be functional and adjusted both to the physical structure and requirements of an elderly person. Unfortunately, as a result of changes occurring in body structure with age, elderly people cannot find appropriate clothes for themselves. In Poland this problem was neglected for a long time, and the world reports that can contribute to the design of clothes for the elderly are scarce and concern mainly fastenings.

Studies of Sperling and Karlsson (1989) from the University of Göteborg confirm the fact that both the ageing process and past diseases significantly restrict motion capabilities, and that the group under investigation requires an exceptionally serious and keen attitude towards the clothing issue. For the elderly, similarly as for the disabled, the question of independent self-service is very important; they do not want to rely on others. There is a direct link

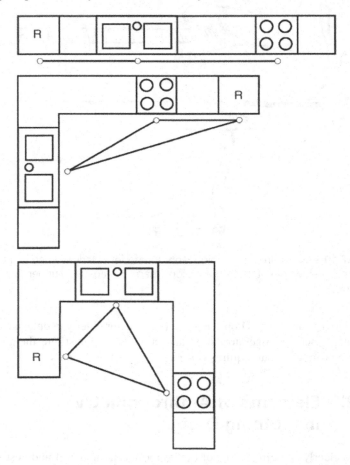

FIGURE 10.4 Lay-out of the kitchen working space—the triangle of work.

between this matter and satisfaction with life, as well as the physical and emotional well-being of an individual. Taking into account the above, the authors of the study made it their primary aim to facilitate the independent living of people who require care, and the possibility to put on and take off clothes was one of the assumptions. The research work, conducted in a systematic way, included: determination of the needs and preferences, analysis of the manual abilities of the population under investigation, development of functional requirements concerning fastenings, and the design and realization of the prototype solutions for such elements meant for elderly people.

Typical anthropometric research for the needs of designing clothes for the elderly was undertaken at the Institute of Industrial Design (IWP) in Warsaw (Nowak *et al.* 2003). The investigation embraced 142 women living in Warsaw who were aged 60–80. Eighteen somatic characteristics were measured, including stature, body mass, crotch height, widths of shoulders, chest and hips, circumferences of neck, trunk, upper and lower extremities, arc of trunk length, and external arcs of extremities. Body mass index (BMI) was also calculated. The results are shown in Table 10.4.

The research results were analysed against the Polish standard PN-84504/1997 for clothing products. It was found that 13% of the women had a stature measurement below the minimum threshold included in the Polish Standard, which amounts to 152 cm. A considerable percentage of the elderly women (76%) were located within the range 152–164 cm, and only 10% were taller and had the stature measurement from 164 to 176 cm. The women subjects were characterized by relatively big chest circumferences. In approximately 60%, this feature measurement was from 96 to 112 cm and 15% had the chest circumference of 112–120 cm. Only 3% of the women had measurements above the range adopted by the Standard. It appeared, however, that 60% of the subjects had the same circumferences of the chest and the hips situated within the range 96–112 cm. The results of the investigation provided data for the design of clothing meant for elderly women. They also showed the necessity to revise existing data used for designing clothes for elderly women and correct the Polish Standard since clothes made according to the data included in the Standard will not suit 13% of women. It is supposed that the percentage would have been bigger if the investigation had covered peasant women, since in Poland there are still significant differences in body measurements between people living in urban and rural areas.

Another anthropometric research work for the needs of clothing design carried out at the IWP embraced a group of elderly men (Nowak *et al.* 2003). It should be mentioned here that all the measurements taken at the IWP for the needs of clothing design had been preceded by the investigation of the users' needs. The investigation was conducted within the target project which was partially financed by the Chief Technical Organizations (NOT) and the manufacturer.

TABLE 10.4 Anthropometric Data of Polish Females Aged 60–80 for the Needs of Clothing Design (in mm)

FEATURE	MEAN DEVIATION	STANDARD VALUE	MINIMUM VALUE	MAXIMUM
Body mass (kg)	68.82	10.86	43.00	95.00
Stature	1570	55	1410	1714
BMI index	27.95	4.47	17.70	39.41
Shoulders width	365	17	331	422
Chest width	276	25	228	345
Hips width	347	22	300	415
Neck circumference	351	22	300	405
Chest circumference	1031	90	810	1230
Waist circumference	888	103	670	1105
Hips circumference	1062	88	890	1300
Thigh circumference	555	57	420	715
Arm circumference	297	33	210	390
Wrist circumference	163	33	140	190
Back length arc	398	29	340	480
Front length arc	537	36	450	650
Neck to wrist length arc	755	35	660	845
External arc of upper extremity length	579	27	525	647
External arc of lower extremity length	1014	45	895	1120
Crotch height	717	42	621	845

The measurement included 144 men from Warsaw aged 65–75. Anthropometric investigation comprised 24 somatic features, including height and width characteristics, arcs, and circumferences. The additional measurement of the arms decline was carried out at the request of the producer. The measurement was taken with the use of a special prototype electronic device designed and made at the Institute of Industrial Design. Somatic characteristics of the men under investigation are presented in Table 10.5. The table includes the means (M), standard deviations (SD), and percentile values of body dimensions.

The results of the investigation were compared with the Polish Standard and showed that the majority of the population under investigation (84%) were men with body height from 158 cm to 176 cm, 83% of the men had a chest circumference of 96–116 cm and 75% had a waist circumference of 86–110 cm. Seventy-six percent of the subjects had a deep chest, 82% had a female figure,

TABLE 10.5 Anthropometric Data of Polish Males Aged 65–75 for the Needs of Clothing Design (in mm)

FEATURE	STANDARD MEAN	5TH DEVIATION	50TH PERCENTILE	95TH PERCENTILE	PERCENTILE
Body mass (kg)	79	13	60	79	104
Stature	1698	60	1595	1699	1790
Height of the cervicale point	1466	55	1377	1464	1553
Perineal height	782	43	705	780	849
Shoulders width	399	17	371	400	420
Chest width	316	18	285	318	343
Chest depth	241	22	208	238	281
Hips width	328	19	299	328	359
Neck circumference	398	26	352	400	435
Chest circumference	1045	82	902	1044	1179
Waist circumference	972	134	772	970	1185
Hips circumference	1016	67	922	1010	1140
Hips circumference with regard to the abdomen convexity	1072	87	951	1060	1257

(Continued)

TABLE 10.5 (Continued)

FEATURE	STANDARD MEAN	5TH DEVIATION	50TH PERCENTILE	95TH PERCENTILE	PERCENTILE
Thigh circumference	535	40	465	540	594
Arm circumference	299	28	256	300	349
Arc of the front length through the chest	547	39	490	550	604
Arc of the back length	419	27	380	420	470
Arc of the shoulders back width	437	24	400	435	479
Arc of the back width at the height of the chest	389	34	340	388	440
Arc of the front width through the chest	445	41	375	450	514
Armpit width	119	12	100	118	138
External arc of the upper extremity Length	621	25	581	620	660
External arc of the lower extremity Length	1090	49	1011	1090	1155
Arms decline (in grades)	21.9	4	15.1	21.6	28.3
BMI Index	27.48	4.0	21.03	27.25	34.49

i.e. comparatively wide hips in relation to the shoulders width. In the group under investigation, as many as 17% were short men with a body height of under 164 cm. Considering the fact that this height constitutes the lower limit of body height for men in the tailors' tables used in Poland, as many as 17% of the subjects are not included in the tables and cannot buy appropriate clothes on the market.

The results of the anthropometric research and the analysis of the requirements were used to develop clothing designs and make prototypes. The prototypes underwent functional tests and finally a collection of garments adjusted to the needs and anatomic structure of men aged over 65 was made by the producer. It should be noted that the above anthropometric research was conducted in conformity with the classical methods used by anthropometry (after the ISO Standard 8559/1989 Garment construction and anthropometric surveys—body dimensions). At present, anthropometry increasingly makes use of the methods which make it possible to take measurements in a two- or three-dimensional system, mainly for the needs of clothing design. One of the methods of this kind which should be preferred while measuring elderly or disabled persons, among others for the reason of the short duration of measurement, is the method developed by the team of the Defence and Civil Institute of Environmental Medicine in Toronto (Meunier and Yin 2000). The system under review is a PC-based system comprised of two Kodak DC120 colour digital cameras (1280 X 960 pixels) and a blue backdrop embedded with calibration markers. The results of the analysis showed that image-based systems can provide anthropometric measurements that are quite comparable to traditional measurement methods performed by skilled anthropometrists, both in terms of accuracy and repeatability. The quality of the results depends, in large part, on the dependability of the automatic landmarking algorithms and the correct modelling of indirect measurements. Once that is achieved, however, this type of system can provide a uniform measurement of a population regardless of where, when, or by whom, it is operated.

ACKNOWLEDGEMENTS

I would like to express my gratitude to my colleagues and friends from the Institute of Industrial Design in Warsaw for their assistance in preparing the material for this work.

I am also grateful to the IWP research teams for their commitment and assistance in the work, the results of which are presented in this chapter.

REFERENCES

BORGES, O., 1989, Isometric and isokinetic knee extension and flexion torque in men and women aged 20–70. *Scandinavian Journal of Rehabilitation Medicine*, 21, 43–53.

COLEMAN, R., 1993, A demographic overview of the aging of first world populations. *Applied Ergonomics*, 24, 5–8.

DELARUE, J., CONSTANS, T., MALVY, D., PRADIGNAC, A., COUET, C. and LAMISSE, F., 1994, Anthropometric values in an elderly French population. *British Journal of Nutrition*, 71, 295–302.

FIDANZA, F., SIMONETTI, M.S., CUCCHIA, M., BALUCCA, G. and LOSITO, G., 1984, Nutritional status of the elderly. *International Journal of Vitamin Nutrition Research*, 54, 75–90.

JAROSZ, E., 1998, *Anthropometric Data of Elderly People for the Needs of Design*, *Prace i Materialy IWP* (Warsaw: Institute of Industrial Design), p. 153 (in Polish).

KIRVESOJA, H., VÄYRYNEN, S. and HÄIKIÖ, A., 2000, Three evaluations of task-surface heights in elderly people's homes. *Applied Ergonomics*, 31, 109–119.

KOTHIYAL, K. and TETTEY, S., 2001, Anthropometry for design for the elderly. *International Journal of Occupational Safety and Ergonomics*, 7(1), 15–34.

MEUNIER, P. and YIN, S., 2000, Performance of a 2D Image-based anthropometric measurement and clothing sizing system. *Applied Ergonomics*, 31, 445–451.

MOLENBROEK, J.F.M., 1987, Anthropometry of elderly people in the Netherlands; research and applications. *Applied Ergonomics*, 18(3), 187–199.

NOWAK, E., 2000, *The Anthropometric Atlas of the Polish Population — Data for Design* (Warsaw: Institute of Industrial Design).

NOWAK, E. and KALKA, E., 2003, Anthropometry for the needs of the elderly and disabled. Data for design. *Prace i Materialy IWP*, 1, CD- ROM.

NOWAK, E., LAPACZEWSKA, K. and KALKA, E., 2003, Ergonomic recommendations for clothing design including the needs of elderly men. In *Proceedings of the IX International Scientific Conference — Ergonomics for the Disabled 'MKEN 2003'*, Lódz, pp. 32–45.

PHEASANT, S.T., 1996, *Bodyspace: Anthropometry, Ergonomics and Design* (London: Taylor and Francis).

RAVAGLIA, G., MORINI, P., FORTI, P., MAIOLI, F., BOSHI, F., BERNARDI, M. and GASBARRINI, G., 1997, Anthropometric characteristics of healthy Italian nonagenarians and centenarians. *British Journal of Nutrition*, 77, 7–17.

SMITH, S., NORRIS, B. and PEEBLES, L., 2000, *Older Adultdata. The Handbook of Measurements and Capabilities of the Older Adult — Data for Design Safety* (Nottingham: Institute for Occupational Ergonomics, University of Nottingham).

SPERLING, L. and KARLSSON, M., 1989, Clothing fasteners for long-term-care patients. *Applied Ergonomics*, 20(2), 97–104.

FURTHER READING

Dianat, I., Molenbroek, J., & Castellucci, H. I. (2018). A review of the methodology and applications of anthropometry in ergonomics and product design. *Ergonomics*, *61*(12), 1696–1720.

Frenzel, A., Binder, H., Walter, N., Wirkner, K., Loeffler, M., & Loeffler-Wirth, H. (2020). The aging human body shape. *NPJ Aging and Mechanisms of Disease*, *6*(1), 5.

Kaewdok, T., Sirisawasd, S., Norkaew, S., & Taptagaporn, S. (2020). Application of anthropometric data for elderly-friendly home and facility design in Thailand. *International Journal of Industrial Ergonomics*, *80*, 103037.

Padilla, C. J., Ferreyro, F. A., & Arnold, W. D. (2021). Anthropometry as a readily accessible health assessment of older adults. *Experimental Gerontology*, *153*, 111464.

Silva, N. D. A., Pedraza, D. F., & Menezes, T. N. D. (2015). Physical performance and its association with anthropometric and body composition variables in the elderly. *Ciencia & Saude Coletiva*, *20*, 3723–3732.

Anthropometry for the Needs of Rehabilitation

11

E. Nowak

11.1 INTRODUCTION

At present, a new direction of ergonomics is appearing aimed at the needs of disabled people. This is *rehabilitation ergonomics* (Kumar 1992; Nowak 1992, 1996), roughly defined as an interdisciplinary field of science that aims at adjusting tools, machines, equipment, and technologies, as well as material work and life environments including objects of daily use and rehabilitation equipment, to the psychophysical needs of the disabled. Rehabilitation ergonomics plays a part both in the rehabilitation process and in equalizing opportunities.

Anthropometry plays an important role in rehabilitation ergonomics. It is defined as a set of measuring methods and techniques that allow the investigation of the differentiation of the measurements of the man and their changeability during ontogenesis and phylogenesis.

As ergonomics develops, and in compliance with its needs, anthropometry develops new methods that can be called ergonomic anthropometry. Many of

DOI: 10.1201/9781003459767-11

these methods, concerning both classical and ergonomic anthropometries, can be applied to the rehabilitation ergonomics that can be divided into two parts:

- The first part is strictly connected with ergonomics, where anthropometry provides data for designing and shaping the work and life environment of the disabled.
- The second part embraces all methods and measuring techniques that assist the rehabilitation process.

11.2 ANTHROPOMETRIC RESEARCH OF THE DISABLED—DATA FOR DESIGNING

11.2.1 Somatic Characteristics

Defining the possibilities and necessities of the disabled is a necessary condition for designing the material work and life environment for this population. Data that characterize the somatic structure constitute basic information.

The influence of disabilities on shaping the body structure has been studied by many anthropologists (Molenbroek 1987; Nowak 1989, 1996; Das and Kozey 1994; Jarosz 1996; Pheasant 1996). The largest disproportions between the healthy and the disabled population can be found in a group of people with motor dysfunction. This is understandable since dysfunction results from past or currently developing diseases that lead to joint disturbances of the osseous, ligament and joint, muscular and nervous systems. These disturbances lead to deformities and somatic changes of particular parts of the body. This then affects the final shape and dimensions of the body and its motorics.

Other factors that restrain the development and growth of the body include the restriction of motion activities, neglected nursing, improper or lack of rehabilitation, as well as stresses connected with pain and frequent stays in hospitals and rehabilitation centers which accompany the pathological process. Descriptions of investigations of people with lower extremities dysfunction are found in the literature. They are usually wheelchair users. It should be realized, however, that this group embraces people with various degrees of motor efficiency limitations. This depends not only on the type and stage of a disease, but also on the time of its appearance. Therefore, researchers dealing with this problem face great difficulties in selecting subjects, and in describing results scientifically. This may be the reason for the small number of studies undertaken in this field. This particularly concerns studies where the results are to provide data for designing.

Tables 11.1 and 11.2 present anthropometric measurements collected on the basis of accessible data published in the ergonomic literature (Laubach *et al.* 1981; Boussena and Davies 1987; Goswami *et al.* 1987; Molenbroek 1987; Nowak 1989; Das and Kozey 1994; Jarosz 1996; Pheasant 1996). Only data comparable in respect of the application of the same research methodology were presented. Unfortunately, many data had to be omitted. This concerns, for example, the upper and lateral reaches. According to these studies, the lateral reach can be measured in relation to the shoulder joint (Goldsmith 1967), to the side of the wheelchair (Floyd 1966 after Pheasant 1996), and to the body axis as a half of the lower extremities span (Nowak 1989; Jarosz 1996). The forward reach can be measured in relation to the shoulder joint (Floyd 1966 after Pheasant 1996), to the front of the trunk (Goldsmith 1967), or to the vertical plane of the seat backrest—Bsd (Nowak 1988, 1989; Jarosz 1996).

Height measurements are taken in relation to various reference bases. This makes it difficult not only for comparison purposes, but also for designers who wish to make use of the investigation results. A synthesis of existing data was carried out for design purposes.

Anthropometric data differ significantly, regardless of the fact that they concern various populations. The diseases that necessitate the use of a wheelchair affect the shape of the body. Pheasant (1996) indicates that the body proportions of wheelchair users resemble those of elderly people aged over 65. This was confirmed by the results of the investigations carried out by Molenbroek (1987).

Anthropometric measurements of the disabled are not only important for the needs of design, but also because of the differences between the disabled and healthy population. The majority of authors indicate that the body structure of disabled men and women differs significantly from the able-bodied population (Nowak 1988, 1989; Samsonowska-Kreczmer 1988; Jarosz 1990, 1996; Das and Kozey 1994; Pheasant 1996).

Most of the measurements of Polish people using a wheelchair show smaller values for the disabled than for healthy people. The differences are significant and amount to 110 mm for the seated stature, up to 113 mm for eye level, up to 126 mm for shoulder height, and up to 57mm for elbow height (Figure 11.11a). As far as arm reach measurements are concerned, the differences amount to 204 mm for the forward arm reach and up to 90 mm for the overhead reach. Similar results were obtained by Nowak (1988, 1989, 2000) by comparing the Polish population of disabled young people (aged 15–18 years) with lower extremities dysfunction against able-bodied people of the same age (Nowak 2000).

TABLE 11.1 Structural Anthropometric Data for Males in Relation to the Seat (in Percentiles)

DIMENSION (MM)	PERCENTILES	AUTHORS					
		BOUSSENA AND DAVIES	JAROSZ	NOWAK	MOLENBROEK	DAS AND KOZEY	GOSWAMI ET AL.
Seated stature	5	824	769	744	761	734	—
	95	962	960	972	919	963	—
Eye height	5	—	667	630	643	496	—
	95	—	857	857	810	717	—
Shoulder height	5	—	495	474	520	468	330
	95	—	682	647	649	676	564
Elbow height	5	177	144	158	168	108	136
	95	269	297	289	289	312	212
Knee height	5	483	468	453	—	—	—
	95	586	605	572	—	—	—
Popliteal height	5	381	383	386	401	—	343
	95	473	513	502	503	—	465
Trunk depth	5	—	180	165	211	198	—
	95	—	340	270	344	281	—
Popliteal depth	5	421	461	435	401	—	356
	95	522	636	555	525	—	447
Shoulder breadth	5	383	353	337	—	354	—
	95	482	425	439	—	439	—
Overhead reach	5	—	1028	1022	828	1072	—
	95	—	1324	1320	1214	1415	—
Reach forward	5	568*	653	668	—	—	—
	95	677*	840	861	—	—	—

*Measured from the acromial point.

TABLE 11.2 Structural Anthropometric Data for Females in Relation to the Seat (in Percentiles)

DIMENSION (MM)	PERCENTILES	AUTHOR				
		BOUSSENA AND DAVIES	JAROSZ	NOWAK	MOLENBROEK	DAS AND KOZEY
Seated stature	5	794	668	708	702	647
	95	912	894	890	858	857
Eye height	5	—	570	592	585	546
	95	—	789	783	763	744
Shoulder height	5	—	433	461	479	423
	95	—	619	592	601	597
Elbow height	5	176	133	139	156	105
	95	266	281	309	270	257
Knee height	5	450	407	442	—	—
	95	539	530	532	—	—
Popliteal height	5	364	315	371	361	143
	95	453	454	462	460	182
Trunk depth	5	—	191	182	219	—
	95	—	315	286	368	—
Popliteal depth	5	418	418	424	405	—
	95	516	571	545	524	—
Shoulder breadth	5	368	310	316	—	291
	95	434	394	410	—	355
Overhead reach	5	—	882	963	733	947
	95	—	1192	1195	1113	1090
Reach forward	5	552*	558	617	—	—
	95	630*	713	768	—	—

*Measured from the acromial point

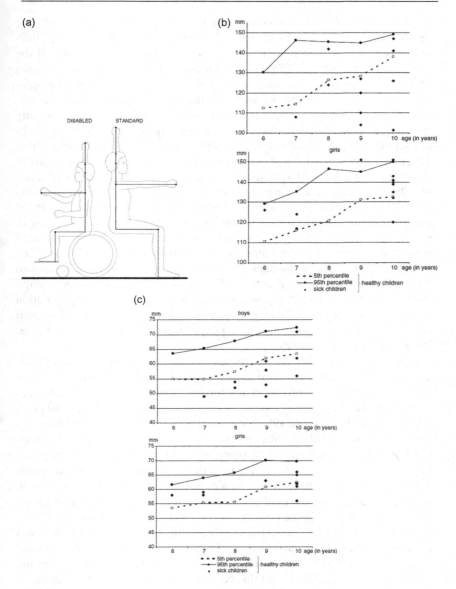

FIGURE 11.1 (a) Comparison of body dimensions of the disabled and able-bodied. (b) Measurements of hand length for children aged six to ten. (c) Measurements of hand width for children aged six to ten.

Floyd (1966) and Bouisset and Moynot (1985) indicate that the smaller seated stature (Bs-v) measurements of the disabled can result from deformities of the osseous system as well as from the fact that as a result of the back muscles paralysis, difficulties with maintaining the straight position of the body appear. This problem seems to be very serious for paraplegics. A similar interpretation can be assumed for the reach measurements. According to Pheasant (1996) the same analogies of the changes in body proportions occur in wheelchair users and elderly people. In old age, there is a similarly as in the case of people with motor organ dysfunction, where deficient muscular tonicity of the chest and abdomen occurs. This results in an increase of pectoral kyphosis. At the same time, the processes of intervertebral cartilages flattening and back shortening occur. This leads to the C shape of the spine (Nowak 1980, 1988). Samsonowska-Kreczmer (1988) indicates that a stooping back and sunken chest occur in young people with motor organ dysfunction. Another reason for the lower values of seated stature measurement is the buttocks and thighs muscles atrophy resulting from the immobility of the back and lower extremities. Studies conducted by Jarosz (1996) indicate that for 55% of men and 65% of women, thigh thickness measurement is below the lower limit of the healthy population standard. Similar results concerning adults were obtained by Goswami et al. (1987), while those in young people were found by Nowak (1988, 1989) and Mięsowicz (1990).

In comparison with the stature and reaches characteristics, the shoulder breadth characteristic (a-a) appears different. Most scientists confirm the fact that the value of this characteristic is larger for wheelchair users than for the healthy population (Boussena and Davies 1987; Goswami et al. 1987; Nowak 1989; Jarosz 1996). Wężyk (1989) and Mięsowicz (1990) point out the fact that children with cerebral palsy have bigger values of shoulder breadth in comparison with those of healthy children. It is supposed that the increase in this characteristic value is caused by the bigger motor activity of the upper extremities. As a comparatively efficient motor organ, the upper extremities are used to carry the whole body while changing the position and displacing the body from a wheelchair to a chair or other piece of furniture. The same concerns moving with the assistance of all kinds of orthopedic aids (such as crutches). Particularly, great activity connected with physical effort is shown in the case of the upper extremities driving the wheels of a wheelchair—in this case they are, in a sense, the propulsion force. Although the development of electronics is changing the method of wheelchair steering, which ensures a disabled person greater comfort, in many countries old generation wheelchairs, requiring the work of muscles, are still in production and use.

Furthermore, the development of the osseous and muscular system of the shoulders is influenced by sports activities of the disabled. Nowadays, this is a widespread and common practice. An additional factor of the development can be the intensive training of these muscles during rehabilitation. It is supposed that the above factors can stimulate the growth of clavicles in their length, the more so as the ossification of the parasternal epiphysis of the clavicles takes place in man last of all in comparison with other long bones. According to Wolan´ski (1983), this process can last until the age of 25 or 27. The stimulating impact of physical training on the increase in bone length was confirmed by Malina (1980) on the basis of the investigations embracing the young. Buskirk *et al.* (1956) also proved that the onesided load of the upper extremity with physical training or work stimulates the increase in its length. It was stated that the bones in the forearm and the hand of professional tennis players were longer in the dominant extremity than in the nondominant one. Prives (1969) noticed lateralization in people practicing sports or performing professional activities loading one side of the body. There are recorded facts of functional lateralization caused by performing one-sided, repeated work (Nowak 1976; Malinowski *et al.* 1985).

Disability occurring in the young age negatively affects further development of the young body and can often lead to various growth disturbances. Young Polish wheelchair users have smaller height and length body measurements in comparison with their healthy peers. In children who became ill with rheumatoid arthritis in their early childhood, apart from advanced changes in the osteoarticular system, growth disturbances and various kinds of local developmental disorders can occur. In approximately 6% of sick children, microsomia prevails and in those who became ill very early, brachycephaly, shortening and narrowing of the cranium segments, as well as atelognathia can occur. An investigation of 25 children aged six to ten years suffering from chronic juvenile rheumatoid arthritis (Nowak *et al.* 2004) indicated that half of the subjects had small body height measurements, located below the lower limit (C5) determined for their healthy peers. Still worse, physical development is noticed in children suffering from infantile cerebral palsy (Wężyk 1989; Mięsowicz 1990; Luczak *et al.* 1993). These children are characterized by short stature, small body mass, narrow hips, deep chest, and weak osseous system. Differences in body structure, in respect of both sizes and directions of the deviations, are similar for males and females. Significant disproportions occur within the trunk area. In both sexes, narrow shoulders and hips (negative normalized values of these characteristics) and, at the same time, broad and deep chest (significantly larger positive deviations of both the transverse and sagittal measurements), were noticed. Narrow hips and large chest dimensions result from the restriction of motor function of the lower extremities for the benefit of the upper limbs, especially when a child uses a wheelchair or other orthopedic aids. It was also observed that differences between the level of the morphological development of sick and healthy children decrease with age. In 16-year-old girls and boys with cerebral palsy shoulder, breadth slightly increases while chest breadth and depth significantly increase.

In the case of blind children the level of physical development is generally worse. They are frail and feeble, their muscular and skeleton apparatus is weak, and they often have faulty posture. They are characterized by a typical body silhouette: hung down head, rounded back, and straight legs (Maszczak 1975). This regularity was confirmed by the results of an investigation carried out on a group of blind children from the Institute for the Blind in Warsaw (Kalka and Cabak 1997). The investigation embraced 98 children aged 6–17 (43 girls and 55 boys). Ten somatic characteristics including stature, body mass, stature in the sitting position, thickness of dermal-adipose folds (on the arm, under the scapular bone and on the abdomen), knee breadth and circumferences (of the arm, hips and shank) were measured. In comparison with healthy peers the blind children appeared to be shorter and lighter with less developed skeleton and musculature. Investigations of deaf children conducted in Poland (Maszczak 1975; Luczak 1993) indicate that hearing impairment also negatively affects the psychophysical development of children. Deaf children are characterized by significantly lower mean values of stature and body mass and smaller hips and extremities circumferences in comparison with healthy children of the same age.

Investigations of Polish children suffering from chronic juvenile rheumatoid arthritis proved that they have significantly smaller dimensions of the hand length and width (Nowak *et al.* 2004) (Figures 11.1b and c).

Rheumatoid arthritis is a disease of the connective tissue and initially leads to the impairment of hand functional efficiency. This is why significantly worse results of the tests concerning the functional characteristics of the hand were recorded for those children in comparison with their healthy peers. This concerned, among others, the force exerted by the hand on a cylindrical handle (Nowak 2004) (Figure 11.2).

Determination of the functional deviations of the hand is important for the needs of designing products meant for people with the hand deformities. The way of handling and manipulating devices and toys should be adjusted to the hand abilities, both in respect of the force exerted and the use of the proper handle shape and diameter. Too small a handle is not adequate for a deformed hand since it cannot grasp the whole of the handle. The lack of the hand–handle contact on the whole palm surface significantly weakens steadiness of the grasp and makes it difficult to perform many manual activities.

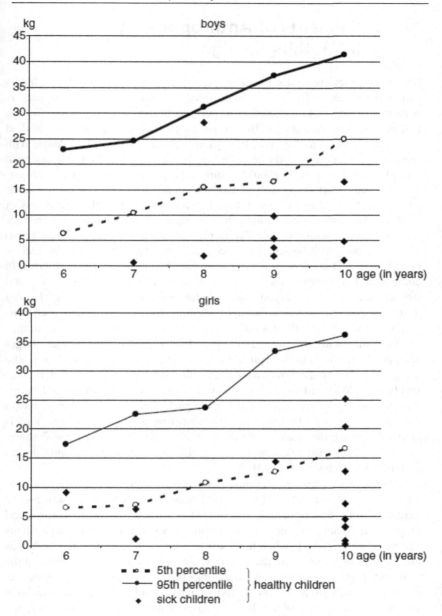

FIGURE 11.2 Force exerted by the hand on a cylindrical handle.

11.2.2　Elements of Anthropometry in Clothing Design

Somatic characteristics are important for clothing design and the fact that its values are bigger in the case of the disabled is of importance for designers. A person suffering from a disability or permanent illness should have clothing that makes their life easier, is functional, meets their emotional needs and, at the same time, is adjusted to the limitations caused by a disability. People with motion dysfunction have the most problems with clothes. Clothing designed for this group of people should assure appropriate physiohygienic properties and heat comfort, as well as ease of manipulation. Concepts of model-constructions solutions should take into account not only the usefulness of products, but also the economic effect. Designs must be simple in construction and easy to manufacture. The form of clothing is to integrate the group of the disabled with society, to conceal a disability, and to give the feeling of satisfaction at possessing clothes appropriate to needs and expectations. In addition to body dimensions, the determination of motor abilities is of vital importance. Based on ergonomic investigations, assumptions for clothing design are determined. For example, clothing for the wheelchair user must fulfill (among others) the following requirements: (1) appropriate clearance allowing the hands to move easily up and forward and appropriately larger shoulder breadth. This is ensured by the special construction of sleeve and armpit cut. (2) Increased transverse dimensions related to the enlarged, muscular chest. (3) Widened sleeve finish related to the considerably enlarged biceps and triceps. (4) Adjustment of the bottom part of clothing length and cut (in the case of a blouse, shirt, and vest) to the sitting position of its user. This is attained through the removal of excess fabric in the front. Figure 11.3 shows selected anthropometric characteristics and their application in clothes design (Dąbrowska-Kielek *et al.* 1993).

People with Down's syndrome (DS) require clothes of a different construction (Harwood 1997). This group of people are always offered oversized garments to fit their large hips and thighs; as a result, all of the features and proportions do not lie in their correct locations. Some of the common problems can be summarized: (1) when the waist of a garment fits, the bust/chest does not; (2) trousers which have the correct waist measurement are too tight on the hips and thighs; (3) dresses and coats that fit the shoulders are too long, i.e. sleeves, trouser legs, and coats are too long; as a result many garment features are at the wrong level of the body; (4) bust contour features are below the bust line, trouser pockets are too low for inserting hands; (5) the full crotch measurement is never long enough to accommodate the over-developed abdomen of the wearers; (6) the overall effect of using standard size garments and

FIGURE 11.3 Utilizing anthropometric data for the construction of clothes meant for the disabled who permanently use a wheelchair (Dąbrowska-Kielek *et al.* 1993).

adapting them to a reasonable fit is obviously unsatisfactory in terms of both their aesthetic and functional characteristics; maternity garments offer a closer approximation to the size requirements and are often used by DS females. The lack of provision is disappointing particularly when block patterns, based on DS body dimensions, are available from which garments could be designed for industrial production.

Many investigators and designers of clothing for the elderly and the disabled underline the fact that in the process of clothing design they are governed by the idea that clothing, in addition to its functional characteristics, should give its user physical comfort. The disabled want to live and work among healthy people. Therefore, they have to accept the clothes they wear. Clothing should not deform the shape of the body, but just the opposite, it should cover anatomical defects.

11.2.3 Workspace Measurement

Essential characteristics exerting an influence on workspace shaping are functional characteristics of the upper extremities, i.e. reaches. Values observed in these characteristics are significantly lower in persons with the lower extremities dysfunction, although their upper extremities are qualified as "efficient." Lower values of reach result not only from lower values of the arm and forearm length, but also from limitations in shoulder and elbow joints. In connection with the above, the disabled have difficulty in performing the movements of abduction and extension (Nowak 1989). Many authors found that, in the case of people suffering from rheumatoid arthritis, workspace of the upper extremity is 7–10% lower if a shoulder joint is constrained and 25–33% lower when elbow joint movements are constrained. It is obvious that disorders of these two joints increase the limitation of the upper extremity movements, and thus the efficiency of the workspace is significantly reduced. This is confirmed by investigations conducted by Nowak (1989) and Jarosz (1996). It should be pointed out that particular values of reaches of the disabled refer to the straight position of the body. Thus, they can be increased through forward and lateral movements of the trunk.

Nowak (1989), based on the Das and Grady's method (1983), developed a simple method of defining workspace for arms. This space was determined for disabled young people with the dysfunction of lower extremities, using the wheelchair.

Figures 11.4a and b show how to determine this space in the sagittal plane (maximum sagittal reach, MSR) and in the transverse plane (maximum transverse reach, MTR). They show the difference for reaches between both

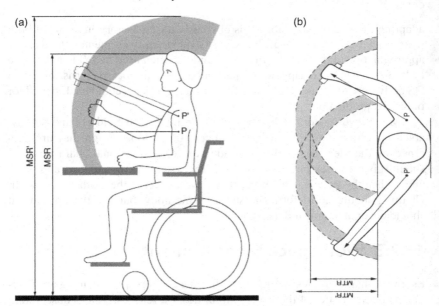

FIGURE 11.4 (a) Maximum sagittal reach for the all-Polish population (MSR′) and the disabled population (MSR). (b) Maximum transverse reach for the all-Polish population (MTR′) and the disabled population (MTR).

disabled and healthy young people. Differences in maximum reach measurements were significant and amounted for the 5th percentile to 300 mm. Results of the study were used in ergonomic analysis at school and in designing school workshops, laboratories, and rehabilitation centers. This method can be recommended for workspace design for the disabled. It is simple and easy to use. Using only five anthropometric characteristics one can obtain a graphic representation of workspace for any population or an individual person.

According to Pheasant (1996), a wheelchair user (whose upper extremities are unimpaired) can reach a zone from ~600 to 1500 mm in a sideways approach, but considerably less "head on." It may well be that the location of fittings within this limited zone will prove entirely acceptable for the ambulant users of the building, but in the case of working surface heights no such easy compromise is possible. As the review of investigations has proved, the body structure of disabled men and women differs considerably from that of the healthy population. Anthropometry, providing data concerning the body structure of disabled people, makes it possible to adjust designs of products and spatial structures to the possibilities and predispositions of this group of users.

11.3 ANTHROPOMETRY FOR THE NEEDS OF REHABILITATION

Anthropometric data and methods can be used for the evaluation of disabled children development and the rehabilitation process of the impaired organ or function.

To check and assess the process of children's physical development, biological reference systems, the so-called standards, are developed. The values of standards are determined on the basis of the population investigations, embracing people qualified as "normally" efficient. Standards are prepared with the use of mathematical and statistical methods. For the needs of pediatrics, two border regions of developmental standards are distinguished:

- The so-called narrow developmental standard, comprising values between the 25th and the 75th percentile and covering 50% of all observations
- The so-called wide standard, comprising values between the 10th and the 90th percentile and embracing 80% of the observations (children included within these limits need observation and concern of a pediatrician)

Developmental standards can also be used to assess the physical development of disabled children. Rehabilitation activities include, among others, the monitoring of somatic development, aimed at the assessment of the child's biological potential. The development of a child depends not only on his genetic potential, but also on environmental conditions.

As a result of disadvantageous events, which can take place during the ontogenesis, injuries inducing disturbances in the nervous system activity, growth or functioning of the whole (or part) body can occur. Biological reactions of the body depend first on the releasing factors, i.e. diseases and personal injuries. These factors have a direct effect on the development of the body, but reactions of the body are also indirectly influenced by other factors, such as the type of medicine and the schedule of its application, as well as the undertaking of rehabilitation. Important factors, which influence the process of development, are the time of disorder occurring and the kind of factor that caused the disorder. Approaching the assessment, one should analyze and determine possible ways of the child's physical development. Developmental standards can be helpful in a single assessment of a child, as well as in continuous investigations; they can also be used in determining the differences between the development of the body structure characteristics of the children with deviation and those of healthy children of the same age. The assessment is usually carried out with the use of the percentile nets representing the height and body mass characteristics. In a single assessment of the child's development one determines the position of the child on the percentile net for the above characteristics and the channel in which this point is located.

Based on the values of the two characteristics, one can determine normality or disorders in the body proportions, for example, the child can have a big body mass (it is in the channel between the 97th and 90th percentile) and low height (between the 10th and the 25th percentile). If one tests the child regularly, the individual curve of their development can be obtained.

Children with motor dysfunction are shorter, they have very narrow hips and laterally flattened chest. They are similar to healthy children in that they have the same shoulder dimensions. They have a low body mass, and this deficiency increases with age.

Anthropometry can be particularly useful in diagnosing and assessing the motor efficiency of the human body. Rheumatic diseases and mechanical injuries result in pathological changes in joints, ligaments, tendons, and muscles and lead to considerably restricted movement ranges. Thus, the assessment of motor efficiency of the affected joints is essential for monitoring the rehabilitation processes.

The simplest way of assessing motor efficiency is by comparing a restricted motion range of an affected joint with its initial range of motion, that is before injury or disease. Unfortunately, after injury or disease has already prevailed, it is impossible to determine what the initial motion range was in the healthy patient. It cannot be expected that a rehabilitated patient had his initial motion ranges measured just before the injury.

Thus, motor efficiency of a rehabilitated patient can only be assessed based upon data of the healthy population. A method based upon this kind of data and allowing the quantitative assessment of rehabilitation progress was developed by Nowak (1992).

Based on the investigation of the ranges of motion of arm, leg, hand, foot, and head standards for the healthy adult population were developed (Nowak 1992). These standards were developed in three motion classes: wide (W), average (A), and small (S). The maximum and minimum movement ranges were developed for particular classes.

Three age categories are subordinated to the classes of movement shown. The youngest persons, aged 18–30, belong to class W, characterized by the maximum values. Class A includes adults aged 31–40, and class S includes the oldest subjects, aged 41–65, whose movement range in particular joints has the smallest values.

Knowing the standard value of particular age classes (φ and the values of movement ranges after the rehabilitation process (φ'), one can calculate the absolute decrease (Ar) and a percentage decrease (Pr) in the movement range. Using a four-grade scale, one can calculate a degree of efficiency.

- Very good: decrease up to 20%—efficiency from 80%
- Good: decrease up to 40%—efficiency from 60%
- Poor: decrease up to 60%—efficiency from 40%
- Very poor: decrease 60%—efficiency 40%

The application of the method above is presented on a graph (Figure 11.5).

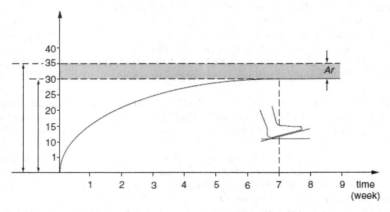

FIGURE 11.5   Example of ankle joint rehabilitation.

ACKNOWLEDGMENTS

I would like to express my gratitude to my colleagues and friends from the Institute of Industrial Design in Warsaw for their assistance in preparing the material for this work. Without their help I would not have been able to prepare the above chapter so quickly and efficiently. I am also grateful to the IWP research teams for their commitment and assistance, the results of which are presented in this chapter.

REFERENCES

BOUISSET, S. and MOYNOT, C., 1985, Are paraplegics handicapped in the execution of a manual task? *Ergonomics*, 28(7), 299–308.

BOUSSENA, M. and DAVIES, B.T., 1987, Engineering anthropometry of employment rehabilitation centre clients. *Applied Ergonomics*, 18(3), 223–228.

BUSKIRK, E.R., ANDERSEN, K.L. and BROZEK, J., 1956, Unilateral activity and bone and muscle development in the forearm. *Research Quarterly*, 27 pp. 127–131.

DAS, B. and GRADY, M., 1983, Industrial workplace layout design. An application of engineering anthropometry. *Ergonomics*, 26(5), 433.

DAS, B. and KOZEY, J., 1994, Structural anthropometry for wheelchair mobile adults. In *Proceedings of the 12th Triennial Congress of International Ergonomics Association — Rehabilitation Ergonomics,* vol. 3, Human Factors Association of Canada, Toronto, 63–65.

DĄBROWSKA-KIELEK, W., SZYMAŃSKA-PETRYKOWSKA, J. and SOCHA-DUDEK, E., 1993, Requirements for clothes design for the disabled using a wheelchair. *Conference "The Human Dimension — Ergonomics and Design"* (Warsaw: Institute of Industrial Design), 68(in Polish).

FLOYD, W.F., GUTTMANN, L., WYCLIFFE-NOBLE, C., PARKES, K.R. and WARD, J., 1966, A study of the space requirements of wheelchair users. *Paraplegia*, May, 1(4), 24–37.

GOLDSMITH, S., 1967, *Designing for the Disabled* (London: RIBA Publications).

GOSWAMI, A., GANGULI, S. and CHATTERJEE, B.B., 1987, Anthropometric characteristics of disabled and normal Indian men. *Ergonomics*, 30(5), 817–823.

HARWOOD, R.J., 1997, The ergonomics of clothing for the disabled and elderly. In *IEA'97 Proceedings of the 13th Triennial Congress of the International Ergonomics Association,* vol. 5, Tampere, Finland, 430.

Wiadomości Instytutu Wzornictwa Przemysłowego, 2, 6-9 (in Polish)

JAROSZ, E., 1996, Determination of the workspace of wheelchair users. *International Journal of Industrial Ergonomics*, 17, 123.

KALKA, E. and CABAK, A., 1997, Physical development of blind children. *Postępy Rehabilitacji*, 11(4), 92–107 (in Polish).

KUMAR, S., 1992, Rehabilitation: An ergonomic dimension. *International Journal of Industrial Ergonomics*, 9(2), 97–108.

LAUBACH, L.L., 1981, Anthropometry of aged male wheel-dependent patients. *Annals of Human Biology*, 8(1), 25–29.

LUCZAK, E., MIESOWICZ, I. and SZCZYGIEL, A., 1993, Somatic characteristics of children and youth with infantile celebral palsy. *Rocznik Naukowy AWF*, Kraków, XXVI, 121–142 (in Polish).

MALINA, R.M., 1980, The influence of physical training on some tissues, dimensions and functions of organism during individual development. *Wychowanie Fizyczne i Sport*, 1, 3–35 (in Polish).

MALINOWSKI, A. and STRZALKO, J., 1985, *Anthropology* (Warsaw Poznań: PWN).

MASZCZAK, T., 1975, Physical development and physical efficiency of deaf children in Poland. *Rocznik Naukowy AWF*, Warsaw, XX, 193–229 (in Polish).

MIESOWICZ, I., 1990, Somatic development of children with deviations. *Rocznik Pedagogiki Specjalnej*, I (in Polish).

MOLENBROEK, J.F.M., 1987, Anthropometry of elderly people in The Netherlands. *Research and Application*, 18.3, 187.

NOWAK, E., 1976, *Determination of the Work Space of the Upper Extremities for the Needs of Workstands Design. Prace i Materialy IWP*, vol. 30 (Warsaw: Institute of Industrial Design) (in Polish).

NOWAK, E., 1980, *Minimum Pressing Force of the Foot for the Needs of Foot-Operated Control Systems. Prace i Materialy IWP*, vol. 55 (Warsaw: Institute of Industrial Design) (in Polish).

NOWAK, E., 1988, *Work Zones for the Disabled. Data for Designing. Prace i Materialy IWP*, vol. 129 (Warsaw: Institute of Industrial Design) (in Polish).

NOWAK, E., 1989, Workspace for disabled people. *Ergonomics*, 9(32), 1077.

NOWAK, E., 1992, Practical application of anthropometric research in rehabilitation. *International Journal of Industrial Ergonomics*, 9, 109.

NOWAK, E., 1996, The role of anthropometry in design of work and life environments of the disabled population. *International Journal of Industrial Ergonomics*, 17, 113.

NOWAK, E., 2000, *Anthropometric Atlas of the Polish Population — Data for Design* (Warsaw: Institute of Industrial Design).

NOWAK, E., 2004, Functional assessment of a child's hand for the needs of ergonomics and rehabilitation. *Ergonomia IJE&HF*, 26(3), 227–252.

PHEASANT, S.T., 1996, *Bodyspace: Anthropometry, Ergonomics and Design* (London: Taylor & Francis).

PRIVES, M.G., 1969, Influence of labour and sport upon skeletal structure in man. *Anatomical Record*, 1, 51–62.

SAMSONOWSKA-KRECZMER, M., 1988, Measurements of disabled youth. Data for clothing design. *Wiadomości Instytutu Wzornictwa Przemysłowego*, vol. 4 (Warsaw: Institute of Industrial Design), (in Polish).

WĘŻYK, E., 1989, Physical development of children and youth with infantile cerebral palsy. *Master's Thesis*, Wroclaw University, Faculty of Anthropology, Wroclaw (in Polish).

WOLAŃSKI, N., 1983, *Biological Development of the Man* (Warsaw: PWN).

FURTHER READING

Bhat, A. K., Jindal, R., & Acharya, A. M. (2021). The influence of ethnic differences based on upper limb anthropometry on grip and pinch strength. *Journal of Clinical Orthopaedics and Trauma, 21*, 101504.

Herrera-Sandate, P., Vega-Morales, D., De-Leon-Ibarra, A. L., Valdes-Torres, P., Chavez-Alvarez, L. A., Hernandez-Galarza, I. D. J., Pineda-Sic, R. & Galarza-Delgado, D. Á. (2021). POS1446 anthropometric measurements in upper extremity rehabilitation of patients with rheumatic diseases. *Annals of the Rheumatic Diseases, 80*:1007, Suppl. 1.

Minetto, M. A., Pietrobelli, A., Busso, C., Bennett, J. P., Ferraris, A., Shepherd, J. A., & Heymsfield, S. B. (2022). Digital anthropometry for body circumference measurements: European phenotypic variations throughout the decades. *Journal of Personalized Medicine, 12*(6), 906.

Wang, T., Zhang, B., Liu, C., Liu, T., Han, Y., Wang, S.,... & Zhang, X. (2022). A review on the rehabilitation exoskeletons for the lower limbs of the elderly and the disabled. *Electronics, 11*(3), 388.

Wicaksono, G. A., Rosa, M. R., & Barri, M. H. (2022, December). Implementation of exoskeleton robots for upper limb rehabilitation based on Indonesian anthropometry. In *2022 2nd International Conference on Intelligent Cybernetics Technology & Applications (ICICyTA)* (pp. 139–143). IEEE.

Index

abdomen 105, 112, 123, 134, 136, 138
abduction 43, 105, 140
acceleration 44, 92–97
access 3, 8, 22
accident 112, 117
accuracy 3–5, 125
acromion 24–27, 29, 48, 100, 115
adduction 43, 105
adult 6, 11, 23, 29, 45, 48, 49, 86, 92, 93, 104, 111, 113, 114, 143
adulthood 92, 114
age 2, 35, 44, 46, 49, 61, 83, 91, 92, 94–99, 102–104, 110–115, 120, 130, 134–136, 142, 143
ageing 110–112, 114, 120
alignment 81, 88–90
allowance 58, 59, 115
anatomy 2, 58
angle 3, 4, 21, 23, 28, 62, 64, 65, 71, 74, 75, 101–103, 112
anthropology 2, 42–44, 50, 83
anthropometer 3, 45, 81
anthropometry 1–7, 14–23, 29, 33, 38, 41–47, 50, 52, 55–58, 65, 67, 70, 72, 75, 77–80, 82, 86, 89–91, 100, 103, 108–111, 114, 119, 125–129, 138, 141, 142; engineering anthropometry 2, 38, 57, 65, 70, 75; ergonomic anthropometry 44–46, 128; structural anthropometry 65
application 4, 7–9, 15, 16, 18, 19, 23–28, 34, 36, 45, 52, 68, 81, 82, 86, 88, 89, 92, 100, 104, 105, 130, 138, 142–146
approach 2, 19, 36, 61, 63, 65, 69, 75, 141
arc 121, 122, 124
arm 6, 15, 24–28, 30, 34, 44, 48, 51–53, 59, 60, 63, 64, 68, 70, 100, 122, 124, 130, 136, 139, 140, 143; upper arm 38, 39, 77
armpit 124, 138
armrest 25
arthritis 135, 136, 140

atlas 49, 106–108
axis 4, 28, 46, 47, 85, 104, 105, 118, 130

back 1, 3, 5, 26–28, 43, 52, 55, 69, 71–75, 82, 101, 113, 115, 122, 124, 134, 136
backrest 27, 101, 102, 117, 130
bend 115, 117
bending 43, 72, 75, 115; bending backward 55; bending forward 55
biacromial 27, 31
biceps 138
bideltoid 27, 31
biomechanics 57, 58, 69, 72, 75–79
biostereometrics 21
body 15–24, 28–35, 42, 47–52, 57–61, 64, 65, 69, 71–74, 82, 83, 86–89, 94–99, 101–103, 105–108, 110, 111, 114, 116–119, 125–129, 131, 132, 134–137, 139, 143
Body Mass Index (BMI) 121, 135, 136, 138
bone 2, 24, 112, 115, 135, 136
breadth 2, 16, 43, 64, 131, 132, 134, 138; buttocks breadth 101; chest breadth 51, 136; foot breadth 28, 31, 51; hand breadth 28, 31, 51; head breadth 28, 31, 51; hip breadth 23, 27, 31, 38, 51, 63; seat breadth 101; shoulder breadth 27, 31, 38, 51, 131, 132, 134, 138
b-spline 17, 84
Bs-v 134
B-sy 95
buttock 27, 101, 134
buttock-knee 27, 30
buttock-popliteal 27, 31
B-v 95

cadaver(s) 21
calibration 3, 125
caliper 3, 16, 81
chair 61, 62, 101, 102, 134
chest 4, 16, 20, 27, 30, 44, 51, 55, 84, 112, 113, 121–124, 134–136, 138, 142
child 6, 20, 91, 101, 102, 104, 136, 142

147

childhood 96, 135
children 6, 9, 45, 50, 91, 92, 94–97, 101–104,
    106–109, 133–136, 141, 142
circumference 16, 43, 44, 55, 136; arm
    circumference 6, 122, 124, 128;
    chest circumference 4, 121–123;
    hip circumference 84
clearance 24, 25, 27, 30, 39, 58, 59, 63, 75,
    138; lateral clearance 24, 62–65,
    75; thigh clearance 60
cloth 52–55, 85, 102, 120, 121, 125, 138, 139
clothing 2, 5, 15, 16, 25, 27, 28, 34, 38, 43,
    55–57, 59, 61, 75, 84, 119–123,
    125, 138, 139
cost 20, 69, 71
Cp 47
CPMS 67, 68
C-shape 101

data 2–4, 39, 44, 47–53, 55–65, 68, 70–75,
    81–83, 85–88, 91–93, 97, 100,
    103, 106–109, 111, 113, 121–123,
    129–132, 139, 141, 143–145;
    anthropometric data 5–10, 16,
    18–22, 23, 33–36, 45, 48, 50, 51,
    58, 59, 74, 75, 81, 86, 88, 91, 92,
    103, 122, 123, 130, 131, 132, 139,
    141
degree 16, 23–27, 47, 92, 129, 143
depth 2, 4, 23, 27, 30, 31, 45, 51, 59, 60, 64,
    66, 101, 113, 123, 131, 132, 136;
    abdominal depth 23, 27, 30; body
    depth 59, 60; buttock-popliteal
    depth 27, 31; chest depth 27, 31,
    51, 123; popliteal depth 27, 31, 51,
    101, 131, 132; seat depth 101; trunk
    depth 131, 132
disability 135, 138
disease 10, 21, 101, 115, 120, 129, 130, 136,
    142
dysfunction 45, 129, 130, 134, 138–140, 142

elbow 24–26, 28, 29, 31, 39, 48, 51, 59–64,
    71, 100–102, 104, 115, 118,
    130–132, 140
elbow-to-elbow 59, 60
elderly 20, 45, 49, 50, 110–121, 125–127, 130,
    134, 139
electromyography 73
ergonomics 2, 5, 7–9, 11, 15, 32–35, 43–45,
    47, 49, 50, 58, 75–80, 90–92, 97,
    111, 112, 114, 118
exoskeleton 7

face 5, 11, 16, 83, 85, 87, 88, 129
finger 2, 24, 26–28, 34
fingertip 24, 26, 28, 29
flexion 43, 72, 104
foot 2, 21, 28, 31, 34, 44, 51, 61, 62, 65, 87,
    88, 104, 143
footprints 15
footrest 44
force 35, 43, 44, 53, 69–74, 76, 82, 94, 134,
    136, 137
forearm 2, 26, 48, 52, 59, 60, 63, 100, 135,
    139
forefinger 58
Frankfurt plane 88
freedom 89

generation 72, 91–94, 134
grasp 136
gravity 1, 117
grip 26, 30, 51, 53, 54, 70

hand 24–26, 28, 31, 34, 38, 44, 48, 51, 53, 54,
    61–63, 70, 71, 73, 84–86, 100, 133,
    135–137, 143
head 2, 21, 24, 25, 28, 31, 38, 43, 44, 48, 51,
    62, 87, 88, 100, 104, 105, 112, 136,
    141, 143
health 6, 10, 11, 28, 34, 60, 70, 78–80, 111,
    113, 127
height 2, 4, 15, 16, 20, 26, 30, 34, 38, 39,
    43, 44, 46–48, 50, 51, 61, 65, 66,
    70, 71, 75, 93–97, 103, 112, 115,
    118, 119, 121–125, 135, 141, 142;
    acromion height 48, 100, 115;
    crotch height 121, 122; elbow
    height 24, 29, 39, 48, 51, 59,
    60, 62, 63, 100, 101, 115, 118,
    130–132; eye height 24, 29, 48,
    51, 59, 60, 62, 64, 75, 100, 131,
    132; fingertip height 24, 29; head
    height 48, 100; hip height 24, 29;
    infrascapular height 101; knee
    height 38, 48, 102, 131, 132;
    knuckle height 24, 29; lumbar
    lordosis height 101; neck height
    48, 100; perineal height 123;
    popliteal height 59, 60, 101, 103,
    115, 131, 132; pubic height (B-sy)
    48, 95, 100; seat height 38, 39,
    101, 115; shoulder height 24, 29,
    51, 59, 60, 63, 64, 130–132; sitting
    elbow height 25, 29, 51, 59, 60;
    sitting eye height 25, 29, 51, 59,

60; sitting height 3, 25, 29, 51; sitting knee height 25, 29; sitting popliteal height 25, 29, 51, 59, 60; sitting shoulder height 25, 29, 51, 59, 60; sitting thigh height 25, 29; suprasternal height 48, 100; table top height 101; truck height 48, 100
hip 18, 23, 24, 27, 29, 31, 38, 44, 51, 55, 63, 64, 84, 104, 105, 112, 113, 121–123, 125, 135, 136, 138, 142
human–environment 37
humanization 5
human–machine 15, 23, 66, 75
human–machine–equipment 34
human-system 7
human–technology 34

impairment 136
index(es) 24, 33, 47, 121, 122, 124
inertia 84, 85
infantometer 3

joint 24, 50, 68, 115, 129, 130, 140, 142, 143

knee 3, 25, 27, 30, 38, 48, 74, 100–102, 104, 113, 115, 131, 132, 136
knuckle 24, 28, 29
kyphosis 112, 113, 134

lateralization 135
leg 24, 25, 104, 143
legroom 38, 39
length 2, 3, 16, 24–26, 28, 30, 31, 34, 38, 43, 45, 48, 51, 52, 55, 59, 60, 63, 64, 88, 92, 96, 100–102, 111, 115, 121, 122, 124, 133, 135, 136, 138, 140; arm length 24–26, 30, 34, 48, 59, 100; foot length 28, 31, 51, 88; forearm length 48, 59, 60, 100; hand length 28, 31, 38, 48, 51, 86, 100, 133, 136
ligament 72, 129, 142
limb 44, 45, 92, 96, 105, 136

machine 45, 52, 53, 58, 86, 128
machinery 7, 8
macro-level 83, 89
male 21, 49–51, 58, 59, 62, 64, 68–70, 89, 104, 113, 114, 123, 131, 135
malnutrition 6
manikin 8, 9, 92, 103–106, 108
manipulation 28, 53, 58, 67, 116, 138

mannequin 67, 75, 78
mass 34, 51, 80, 82–84, 92, 93, 96, 113, 114, 121–123, 135, 136, 142
measurement 1–7, 10–19, 22–25, 31–36, 39, 42–51, 53–55, 58, 59, 61, 63–65, 67–69, 71, 75–85, 89, 93, 94, 100–103, 106, 112, 113, 115, 118, 121, 122, 125, 130, 133–135, 138, 139, 141
method 3, 4, 45, 60, 72, 73, 82–88, 102, 125, 134, 140, 141, 143
microsomia 135
microtrauma 71
middle-aged 113, 115
model 5, 9, 17–19, 21, 33, 36–38, 63, 66–68, 72–76, 81, 84–87, 89, 103–105, 108
motion 14, 18, 24–28, 33, 63, 68, 72, 89, 104, 120, 129, 138, 142, 143
movement 55, 58, 61–63, 69, 73, 74, 104, 113, 142, 143
muscle 27, 69, 70, 72–74, 76, 92, 101, 114, 134, 135, 142
musculoskeletal disorders 9, 57, 58, 69, 73, 74, 76

neck 4, 44, 48, 55, 66, 92, 100, 104, 112, 121–123
newborn 92
nutrition 6, 10, 15

ontogenesis 42, 110, 128, 142
ontogeny 103, 106
organ 18, 21, 83, 101, 103, 110, 114, 134, 141
organism 21
over-exertion 71

palm 26, 136
percentile 20, 29, 30, 38–40, 46–48, 50, 58–60, 62–66, 68, 85, 86, 88, 89, 93, 96, 100, 104, 115, 122–124, 131, 132, 141, 142
plane 43–45, 52, 53, 63, 88, 102, 104, 105, 117–119, 130, 140; frontal plane 52, 117; sagittal plane 43, 45, 104, 105, 140; transverse plane 43, 52, 104, 105, 140
population 1, 6, 7, 16, 19–21, 23, 24, 26–28, 32–36, 42–51, 58, 61–64, 71, 73, 81, 83–86, 89, 92–95, 97, 106–108, 111–113, 115, 117, 121, 122, 125, 129, 130, 134, 140, 141, 143

position 20, 45, 55, 61, 63, 64, 67, 100–104, 115–119, 134, 136, 138, 140, 142
posture 58–64, 66, 70–72, 74–76, 91, 100, 113, 136
puberty 44, 93, 102, 104
pubescence 94

range 7, 9, 11, 19–21, 35, 36, 38, 49, 52, 53, 55, 61, 69, 96, 113, 121, 142–144
reach 15, 21, 26, 27, 30, 48, 51–53, 55, 58, 59, 61–64, 66–68, 70, 71, 74, 75, 97, 100, 104, 105, 115–118, 130–132, 134, 139–141; arm reach 62, 65, 67, 82, 114

scanner 4, 5, 65, 66, 82, 83
seat 25, 27, 38, 39, 45, 52, 101–103, 115, 118, 130–132
segment 1, 20, 38, 44, 45, 57, 58, 69, 73, 102, 104, 135
shape 5, 9, 14, 15, 21, 33–35, 45, 53, 75, 79–92, 105, 112–114, 118, 129, 130, 134, 136, 139
shoe 24–26, 28, 39, 59, 61, 75, 85, 102, 115
shoulder 16, 24–27, 29, 31, 38, 51, 55, 59, 60, 63, 64, 68, 104, 105, 115, 121–125, 130–132, 134–136, 138, 140, 142
span 15, 27, 31, 111, 130; elbow span 28, 31, 101
strength 9, 14, 15, 19, 26, 28, 33, 56–58, 61, 62, 69–71, 73–76, 93; isokinetic strength 69; isometric strength 69
structure 25, 44, 45, 50, 58, 67, 71, 85, 86, 92, 111, 113–115, 120, 125, 129, 130, 135, 141, 142

table 3, 7–9, 23–30, 39, 40, 47–51, 59–64, 69, 70, 74, 75, 85, 95, 97–102, 113–116, 118, 119, 121–124, 130, 131, 132
task 15, 18, 19, 35, 43, 47, 53, 58, 60–64, 66, 69–71, 73, 74, 76, 111, 114
technique 5, 6, 43, 87
thickness 51, 82, 101, 118, 134, 136
thigh 24, 25, 27, 30, 44, 48, 51, 55, 59, 60, 100–102, 105, 118, 122, 124, 134
thora-columbar 104

vertebrae 73, 76
virtual-reality 21

weight 6, 20, 23, 28, 31, 34, 63, 70, 72–74, 76, 84, 86, 101, 113, 114
wellbeing 58, 75, 121
wheelchair 55, 115, 116, 117, 129, 130, 134, 135, 136, 138, 139, 140, 141
width 16, 27, 47, 66, 121, 124, 125; armpit width 123; arm width 15; chest width 122, 123; extended arm width 15; hand width 133, 136; head width 2; hips width 123; seat width 27; shoulders width 122, 123, 125; waist width 4
wrinkles 84
wrist 28, 55, 122

young 45, 91–93, 96, 97, 100–102, 106–109, 111, 113, 130, 134, 135, 140, 141
youth 101

zone 45, 52, 87, 115–118, 141

Printed in the United States
by Baker & Taylor Publisher Services